Open Computing Guide to Mosaic

Levi Reiss &
Joseph Radin

Osborne **McGraw-Hill**

Berkeley New York St. Louis San Francisco
Auckland Bogotá Hamburg London Madrid
Mexico City Milan Montreal New Delhi Panama City
Paris São Paulo Singapore Sydney Tokyo Toronto

Osborne **McGraw-Hill**
2600 Tenth Street
Berkeley, California 94710
U.S.A.

For information on software, translations, or book distributors outside of the U.S.A., please write to Osborne **McGraw-Hill** at the above address.

Open Computing Guide to Mosaic

1234567890 DOC 998765

ISBN 0-07-882088-X

Publisher
Lawrence Levitsky

Acquisitions Editor
Jeff Pepper

Technical Editor
John Navarra

Copy Editor
Deborah Craig

Proofreader
Pat Mannion

Computer Designer
Jani Beckwith

Illustrator
Marla Shelasky

Series Designer
Jani Beckwith

Quality Control Specialist
Joe Scuderi

Cover Designer
EM DESIGN

We dedicate this book to our wives and children:
Noga, Sami, and Maya Reiss,
and
Sara, Gal, and Nurit Radin.

About the Authors

Levi Reiss has been working with computers since 1972 and has designed training materials and taught on subjects such as computer information systems, operating system architecture, systems analysis and design, telecommunications, and programming languages. He has published several textbooks and holds a masters degree in computer science from the University of Paris.

Joseph Radin, currently at DEC, is a senior software specialist and a Mosaic expert. He holds a master of science degree from the Technion, Israel Institute of Technology, which he received upon completing his thesis at the University of California, Berkeley. He and Levi Reiss have collaborated on *X Window Inside and Out* and the *Unix System Administration Guide*. They also authored part of the *LAN Times Guide to Interoperability*.

Contents At A Glance

1 The Internet 1
2 Getting Started with the Windows Version of Mosaic 21
3 Getting Started with X Window System (Unix) Mosaic 39
4 A Guided Tour 55
5 Searching for Information with Mosaic 81
6 The Hypertext Markup Language (HTML) 113
7 Customizing Mosaic 143
8 Mosaic Menus and Buttons 169
A Useful Information Sources 191
B Installing Mosaic 205
C The Server 233

Contents

ACKNOWLEDGMENTS, XIII
PREFACE, XV

1 The Internet 1

Introducing the Internet 3
A Brief History 3
What the Internet Does 4
E-Mail 6
FTP Archives 6
Newsgroups 7
The telnet Command 7
Archie 7
Gopher 8
World Wide Web 8
Some Business Applications for the Internet 10
What You Need 11
TCP/IP 12
SLIP and PPP 13
Internet Addresses 13
Internet Management Issues 15
Billing 15

Security 15
Access 16
Privacy 16
Introducing Mosaic 17

2 Getting Started with the Windows Version of Mosaic 21
Installing and Using Mosaic 22
Connecting to the Internet 22
Testing NCSA Mosaic 23
The Mosaic Main Menu 24
Mosaic for Windows Main Menu 24

3 Getting Started with X Window System (Unix) Mosaic 39
Installing and Using Mosaic 40
Connecting to the Internet 40
Testing NCSA Mosaic 40
The Mosaic Main Menu 41
Mosaic for Unix Main Menu 42

4 A Guided Tour 55
Preliminary Steps 56
Adding a Point of Interest 58
The Radio 59
The Collision of '94 60
What's New with Mosaic 64
Golfing 66
Disabling Inline Images 68
In and Out of the White House 68
Weathering a Hurricane—Looking at Images and Movies 72
Modifying the Screen Presentation 73
Processing Movies and Special Images 76

5 Searching for Information with Mosaic 81
Two Important Mosaic Acronyms 82
Uniform Resource Locator (URL) 82
Hypertext Markup Language (HTML) 84
Different Types of Home Pages 84
Personal Home Pages 86
Corporate and Educational Home Pages 91
Searching Techniques 96
Starting Points 96
Hotlists 97
The History Feature 97
Gopher Servers 99
Applying the Search Techniques 99

Structured Search Techniques 99
Unstructured Search Techniques 103

6 **The Hypertext Markup Language (HTML)** **113**

HTML Elements 114
HTML Structure 115
Elements Describing the Entire Document 116
Text Formatting Elements 116
Embedded Images 119
More Information on HTML 120
Creating HTML Documents with HoTMetaL 120
Downloading the HoTMetaL Software 121
Installing HoTMetaL 122
Overview of HoTMetaL Menus 126
The File Menu 126
The Edit Menu 130
The View Menu 132
The Markup Menu 134
The Help Menu 136
The Window Menu 138
Home Page Creation 138

7 **Customizing Mosaic** **143**

Configuring Mosaic for Windows **144**
Syntax 144
Main Section 145
Settings Section 152
Main Window Section 152
Mail Section 152
Services Section 153
Viewers Section 153
Suffixes Section 155
Annotations Section 156
User Menu Sections 157
Hotlist Section 159
Document Caching Section 161
Font Sections 161
Proxy Information Section 162
Online Documentation 162
Configuring Mosaic for X Window (Unix) 163
Mosaic Information Files 165
Troubleshooting 166
Failed DNS Lookup 166
No Menus 166
HT Access Error 167

8 Mosaic Menus and Buttons 169
Mosaic Menus 169
Mosaic Main Menu for Windows 170
Mosaic Main Menu for X Window (Unix) 180
Mosaic Buttons 184
Mosaic Toolbar for Windows 185
Mosaic Buttons for X Window (Unix) 187
A Useful Information Sources 191
Compendium of Information Sources Appearing in Text 192
Chapter 1 192
Chapter 2 193
Chapter 3 193
Chapter 4 193
Chapter 5 194
Chapter 6 196
Chapter 7 196
Appendix B 196
Gopher Servers 197
B Installing Mosaic 205
Using Internet Chameleon 206
Requirements Checklist 207
Installation 208
Using Dial-up Connections 208
Accessing the FTP Application 209
Trumpet Winsock 214
Installation 215
Configuration 215
Logging into the Server 216
Internet Provider Setup Program 217
Installing the MS-Windows Version of Mosaic 217
Requirements Checklist 218
Obtaining the Software 219
Installing the X Window (Unix) Version of Mosaic 223
Obtaining the Software 223
Customizing Your NCSA Mosaic 226
Viewers and Players 226
Default Settings 226
Finding Viewers 227
Installing MS-Windows Viewers 228
Installing Unix Viewers 229

C The Server **233**
MS-Windows Server Features 234
Installing the Server 234
Downloading the MS-Windows Version 234
Downloading the Unix Version 236
Configuring the Server 238
Starting the Server 247
Shutting Down the Server 248
Shutting Down the MS-Windows Server 250
Shutting Down the Unix System Server 250
Index **253**

Acknowledgments

A computer book is never the sole product of one or two authors, and the *Open Computing Guide to Mosaic* is no exception. We want to thank the outstanding Osborne/McGraw-Hill team for their tireless efforts culminating in this text. Jeff Pepper, the Editor-in-Chief, played a crucial role, signing us and directing the project from its inception to the final stages. We would like to thank him for taking the time and effort to edit every chapter in this text. A lot of what you don't see is due to Jeff. Special thanks go to the Project Editor, Claire Splan, and the copyeditor, Deborah Craig. Picky, picky, picky. It was sometimes frustrating, but in the long run, you, the reader, benefit from their attention to detail. We reiterate our thanks to Claire and Deborah. John Navarra was our technical editor. We want to thank him for his many minor, and occasionally major, corrections. Our heartfelt appreciation goes to the design team for a beautiful, highly functional text designed for maximum pedagogy. We appreciate the support received from David Bradshaw of HookUp Communications and Robert Williams of NetManage. Special thanks go to the Mosaic developers for their outstanding product; without them, this book would not exist. Final appreciation goes to our wives and children—Noga, Sami, and Maya Reiss, and Sara, Gal, and Nurit Radin—for their infinite patience. Hopefully, normal life has returned by the time you read these lines. As is customary, the authors acknowledge their sole responsibility for any errors.

Preface

A recent issue of *ComputerWorld* interviewed the developers of VisiCalc, the first killer app, the one that made the Apple II Plus the choice of the business world. In the article, the question was posed: Is Mosaic the next killer app? We think so.

Not long ago, *ComputerWorld* estimated one million users with a growth rate of thirty thousand per month. The present estimate is more than two million copies of Mosaic in use with another thirty to fifty thousand copies downloaded each month. Mosaic provides access to about three thousand graphical, multimedia databases.

No cheerleading is necessary; this product is hot.

What is Mosaic? Mosaic is a graphical user interface designed for information discovery and retrieval over the global Internet. It enables those without time or patience for convoluted syntax to access desired information, independent of its physical location.

Because people don't always know exactly what they want when they start looking, Mosaic uses sophisticated techniques (often called hypertext) to link from one topic to the next. Within minutes you can cross the globe several times, for example, going from the U.S. to Switzerland to Hong Kong and back to the U.S., gathering information every step of the way. This information can include graphical images, sound, and motion pictures.

Users like you, at home and in the office, will link as Mosaic clients to a wide variety of servers. By applying the information just in the early chapters of our book, you'll be able to install Mosaic and move onto the fast lane of the information highway.

Mosaic applies the client-server model; in this case, the clients retrieve information stored on the server via the Internet.

Some of the many unique features of this book are: helpful tips from a seasoned practitioner; focus on the Mosaic client, without neglecting the WWW server; extensive screen dumps and examples; and thorough installation procedures.

This book includes eight chapters plus three appendixes. We know that you may not have the time to read the book from cover to cover. Begin by reading the brief chapter descriptions below. Then roll up your sleeves and start with the chapter or appendix you want.

Chapter 1: The Internet

This chapter introduces the Internet. Instead of simply telling you how wonderful the Internet is, the chapter discusses problems associated with the Internet. It concludes with an introduction to Mosaic.

Chapter 2: Getting Started With the Windows Version of Mosaic

Chapter 2 provides a running start to MS-Windows Mosaic. (For installation instructions, go directly to Appendix B.) You'll get an introduction to Mosaic menus and the toolbar. When you finish this chapter, you may proceed to Chapter 4.

Chapter 3: Getting Started With X Window System (Unix) Mosaic

Chapter 3 gets you started with X Window System (Unix) Mosaic. (For installation instructions, go directly to Appendix B.) You'll get an introduction to Mosaic menus and the toolbar.

Chapter 4: A Guided Tour

The title says it all. This chapter explores Mosaic using applications as varied as viewing the collision of the epoch (when a comet crossed the path of the planet Jupiter), exploring golf links, checking the President's schedule, and watching a hurricane movie.

Chapter 5: Searching For Information With Mosaic

Now you are ready to apply specific techniques to optimize your Mosaic search. First you'll learn some indispensable Mosaic terms and concepts. Then you'll apply structured and unstructured search techniques to help you find what you want, painlessly and effortlessly (or almost).

Chapter 6: The Hypertext Markup Language (HTML)

This chapter presents the language used in creating Mosaic documents such as your Home Page, a startup document. It introduces software that simplifies and speeds the document creation process.

Chapter 7: Customizing Mosaic

Chapter 7 shows you how to configure MS-Windows and Unix Mosaic for maximum performance and ease of use. If you plan to use Mosaic on a regular basis, apply the concepts and techniques contained in this chapter. You won't regret the investment of time.

Chapter 8: Mosaic Menus and Buttons

This chapter presents in detail the menus and buttons that appear in MS-Windows and X Window (Unix) Mosaic. Learn how to use the menus and buttons that you need, and you'll save precious time and energy.

Appendix A: Useful Information Sources

Appendix A provides the addresses of all information sources used within this text. It contains an extensive list of additional information sources accessible from Mosaic.

Appendix B: Installing Mosaic

This appendix tells you how to get the MS-Windows and X Window (Unix) versions of Mosaic up and running. It provides installation procedures for two commonly used network software packages, Chameleon and Trumpet.

Appendix C: The Server

Appendix C introduces the whys and wherefores of the WWW server providing information to Mosaic clients. It shows you how to install and start the server, and how to shut it down.

CHAPTER 1

The Internet

Mosaic is a graphical user interface that makes it easy for you to track down and retrieve information over the Internet. In addition to text, Mosaic can handle graphics, sound, and movie data. To use Mosaic, you simply point to and click on topics of interest. This can lead to other topics of interest, which can lead to other topics, and so forth. The best thing about Mosaic is that it lets you make your way around the Internet without having to remember arcane command sequences. Mosaic was developed for people like you: people who don't have the time or patience to memorize convoluted syntax but need information contained in the Internet, and need it fast.

When you begin searching for information, you don't always know precisely what you're looking for. For this reason, Mosaic uses sophisticated techniques to link one topic to the next. Once you've learned a few basics about how to use Mosaic, you'll be able to glide seamlessly from one subject to another, stopping to take a closer look at topics you find interesting.

The first version of Mosaic, developed to run on Unix-based X Window systems, was released in April 1993. The first version running on MS-Windows systems was released in the Fall of 1993. And the Internet hasn't been the same since. A major computer publication recently estimated Mosaic users at two million, with a growth rate of thirty to fifty thousand per month. Is Mosaic the next killer app, the one you absolutely must have? We certainly think so. Figure 1-1 shows an introductory Mosaic screen. Compare that to the text-based Internet screen in Figure 1-2. Which one would you rather use?

Users like you, armed with microcomputers running Mosaic, can gain access to Internet data anywhere in the world. This text explains how to install Mosaic and move into the fast lane of the information highway. Before exploring Mosaic, however, you should know something about the basics of the Internet and the previous generation of tools for Internet access.

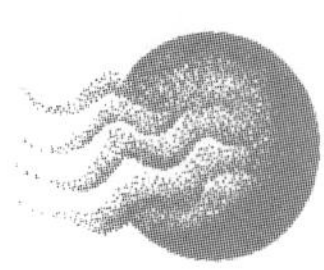

NOTE

This book assumes that you are not a dolt, a dunderhead, or a dummy. At the same time, you are assumed to have intermediate-level knowledge of DOS or Windows. You'll also retain more information if you try the exercises armed with a functioning computer.

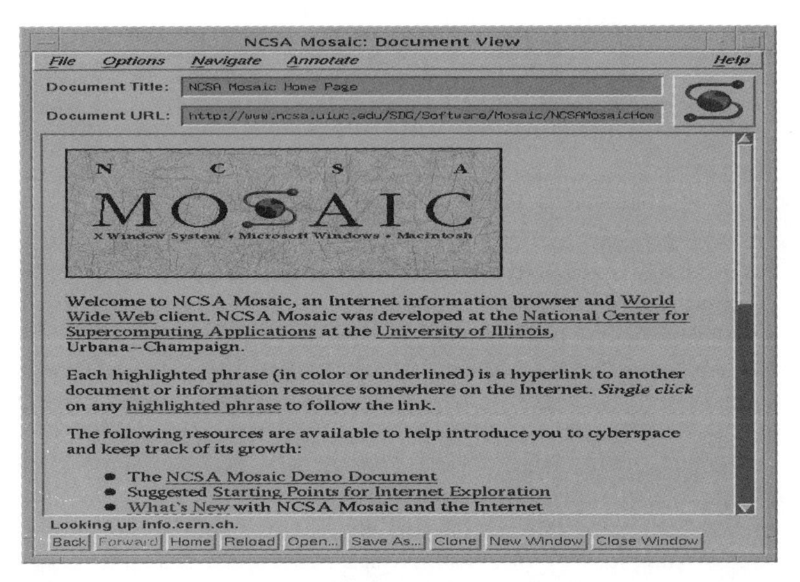

FIGURE 1-1. *An NCSA Mosaic introductory screen*

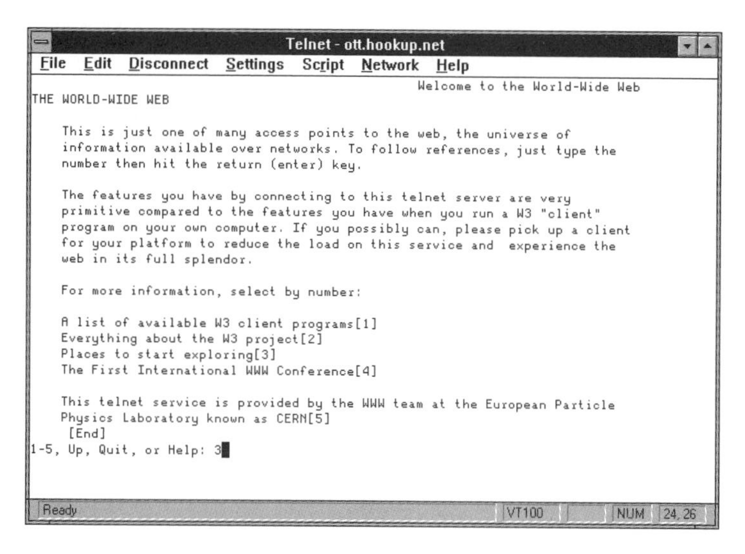

FIGURE 1-2. *A text-based Internet screen*

Introducing the Internet

While any estimate of the Internet's size and scope is greatly exceeded before the ink dries, at last count it included 15 million users, 2 million computers, and 21,000 computer networks. If you don't know what the Internet is, read this chapter to find out what it is and what it can do for you. Even if you are familiar with the Internet, take a few minutes to skim this chapter. Who knows, you may even learn something new.

A Brief History

In case you're wondering what the Internet actually is, it's essentially a huge number of worldwide computers linked together—a network of networks, in other words. Launched as a nationwide computer network for scientists and the military, the Internet dates back to the late 1960s, when it was called the ARPAnet, in honor of its sponsor, the Advanced Research Projects Agency of the U. S. Defense Department. The first successful trial connected four research computers in California and Utah. This is a long way from the estimated 21,000 networks presently linked, a figure that increases by almost 100 percent per year. While scientists and researchers continue to make heavy use of the Internet, business people, students, and computer novices to specialists are also rapidly becoming Internet devotees.

What the Internet Does

Think of the Internet as a library. What a library! This library is open 24 hours a day, 7 days a week. It spans continents. Provided you have the right "library card," you can read (and often copy) a virtually unlimited set of materials, including books, magazines, videos, and X-rays. The sky is not the limit; authorized parties can access communication satellites to get the information they need. The following examples illustrate a tiny portion of the limitless applications for the Internet.

- Billions of people watched World Cup USA '94. These fans can gain access to information on any of the games or teams via the Internet, as shown in Figure 1-3.
- For most people, buying a home is the largest financial transaction of a lifetime. The Internet provides a wide range of information about home buying, including lists of homes for sale and mortgage conditions, as shown in Figure 1-4.
- After seeing the World Cup, you may want to visit the game sites. Or if your team did poorly, a cruise might be just what the doctor ordered. Figure 1-5 shows sample travel information available on the Internet.
- Some companies spice up their information services with a wide variety of information. An example, provided by DEC, is shown in Figure 1-6. Note the Future Fantasy Book Store and information about the beautiful city of Palo Alto, California.

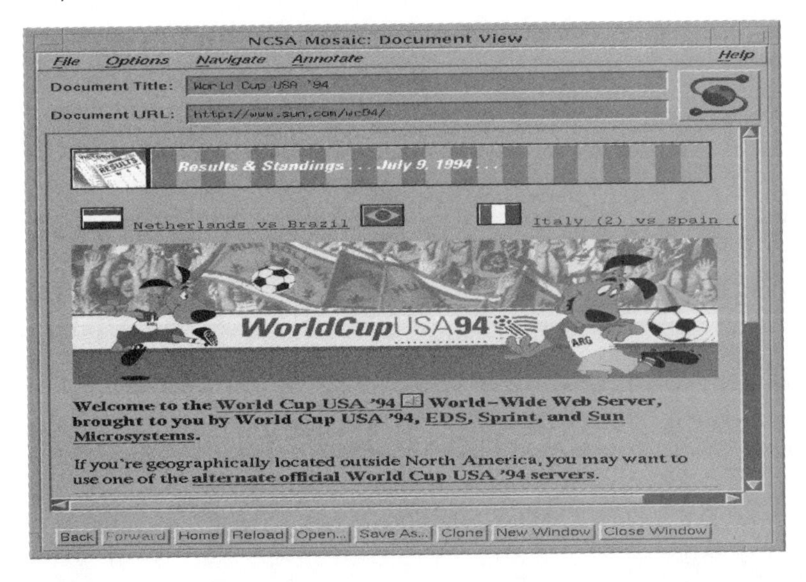

FIGURE 1-3. *World Cup USA '94*

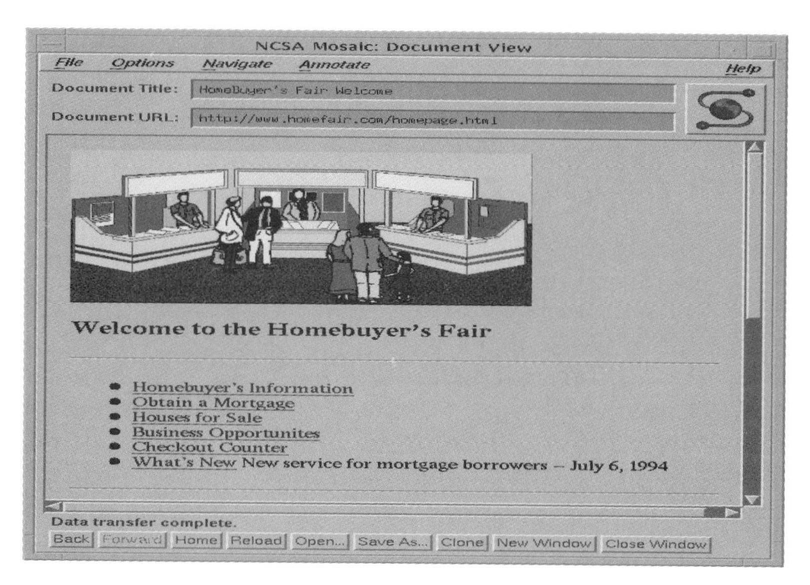

FIGURE 1-4. *Buying a home*

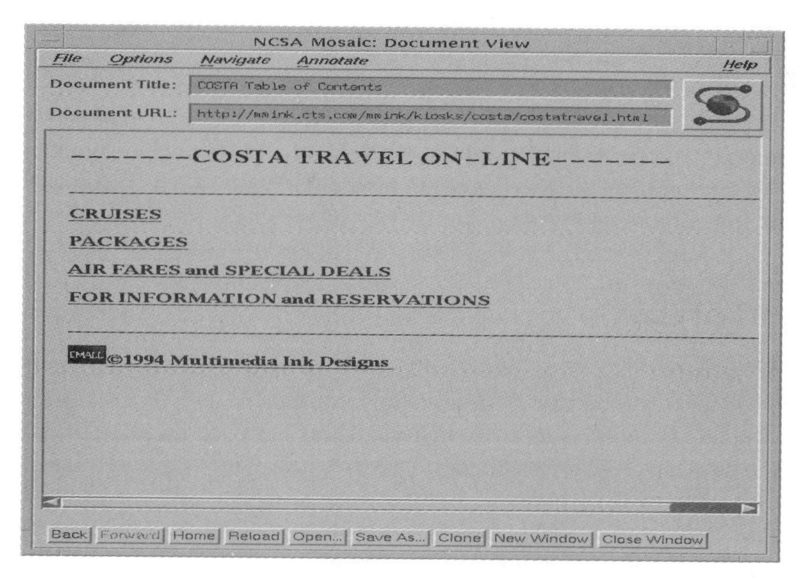

FIGURE 1-5. *Traveling*

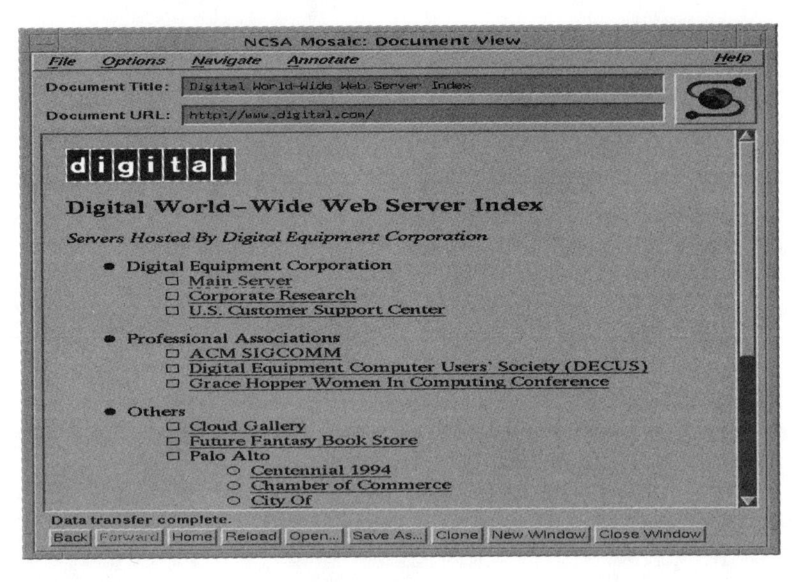

FIGURE 1-6. *The world according to DEC*

E-Mail

Electronic mail, or *e-mail,* is the ability to send and receive messages using your computer. E-mail can be found virtually everywhere: on internal company networks and on the Internet. Literally millions of people employ e-mail to send and receive messages and files, query databases, and gain access to Internet services.

Standard electronic mail can only handle plain text files (technically known as *ASCII files*) and perhaps *binary files* containing other data such as compiled programs and compressed files.

FTP Archives

FTP archives are publicly accessible computers containing files on the Internet. In essence, whatever can be stored on disk can be placed in an FTP archive. You can use the **ftp** (file transfer protocol) command to gain access to these archives. To reduce the burden of assigning individual users a password for the remote computer system, most commercial Internet sites allow file transfers via the anonymous user name. Instead of a password, users may be asked to furnish information such as their electronic mail address.

Newsgroups

USENET News, also called *Net News,* has its origins in an electronic bulletin board set up by students at Duke University and the University of North Carolina. Unlike television and radio news services, Net News is interactive for the most part. Subscribers, and there are millions, can comment on articles of choice. Intercontinental debates rage on multiple subjects.

Given the volume of available news, subscribers must choose one or more areas of interest, or *newsgroups,* of which there are now approximately 4,000. Newgroups are really like discussion groups with topics covering far more than just current affairs. To read the news you must have access to an Internet host and a special program called a *news reader.* The news reader handles newsgroup subscriptions and cancellations, and keeps track of the articles you read and your responses. Sophisticated news readers organize news articles into *threads,* a news article and all responses to it. Before responding to an article, you should read the associated thread. Your pearl of wisdom may already have been offered by others. Information about news readers is found in **news.software.readers.**

As you move from left to right in newsgroup names, the categories become more specific. The newsgroup **news** contains information about USENET. The newsgroup **news.software** contains information about software for USENET. And the fictitious newsgroup **news.software.readers.lefties** contains information about news readers for the left-handed.

The telnet Command

The **telnet** command allows you to gain access to a remote machine on the Internet, provided you have a valid user name and password. Some Internet databases are only available via telnet. However, you may be able to access remote databases without the overhead associated with a telnet login by using FTP archives.

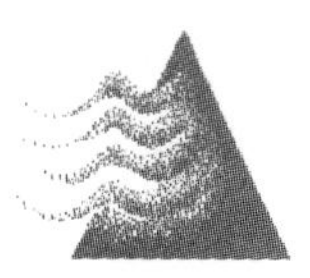

TIP
Don't forget to logoff from the remote computer to avoid tying up valuable resources.

Archie

Archie, developed at McGill University in Montreal, Canada, performs key word searches of titles in Internet databases via anonymous ftp. On a regular basis, Archie searches hundreds of ftp sites and publishes a list of available information. You can query this list via Archie or electronic mail, specifying key words that you want to track down. You can login to an Archie server by using the command **telnet** and providing the login name archie.

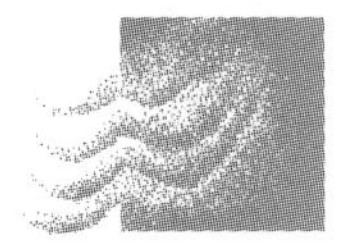

CAUTION
A poorly worded information request may generate a huge number of responses. For example, if you search for the key word "Internet" you'll probably get hundreds of matches, but if you search for something like "baroque music" or "hip hop," you'll probably get a more reasonable number of responses.

Gopher

Gopher is a menu-driven Internet browser that was developed at the University of Minnesota in 1991. Gopher was designed to work with the Internet, rather than merely being tacked on to it. This increases somewhat both system efficiency and ease of use. Presently over 7,000 Gopher centralized information sites (*servers*) are in operation, with more coming online almost every day. A typical request for information may involve several Gopher servers.

Several functions make Gopher easier to use. *Bookmarks* allow you to access a desired menu item without having to step through prior menus. *Veronica* is the short form for *Very Easy Rodent-Oriented Net-wide Index to Computer Archives.* As you might guess from its name, Veronica is similar to Archie. It maintains a database of menu item titles accessible from a Gopher server linked to the Mother Gopher at the University of Minnesota. *Index searches* let you find the desired information rapidly, if it is indexed by keyword. Figure 1-7 shows a Gopher menu for a Canadian weather forecasting application.

World Wide Web

Menu-driven applications such as Gopher can be visualized as trees. As you step through the menus, you go from the major branch (often called the *root*—sorry, the analogy is not perfect), to progressively thinner branches, and finally to a leaf. This type of information search often works well. However, if you don't find the desired information at a given leaf, you must backtrack to the root and try again. The process can be frustrating and time consuming.

An alternative information structure is a *web*, like a spider web, in which you can go from any location to any other location without backtracking to the root. A web structure provides denser connections between points than does the tree structure. It may be harder to set up and maintain, but is definitely more efficient for extracting information. On the Internet this structure is implemented in the *World Wide Web*, often called *WWW*, *W3*, or simply *the Web.*

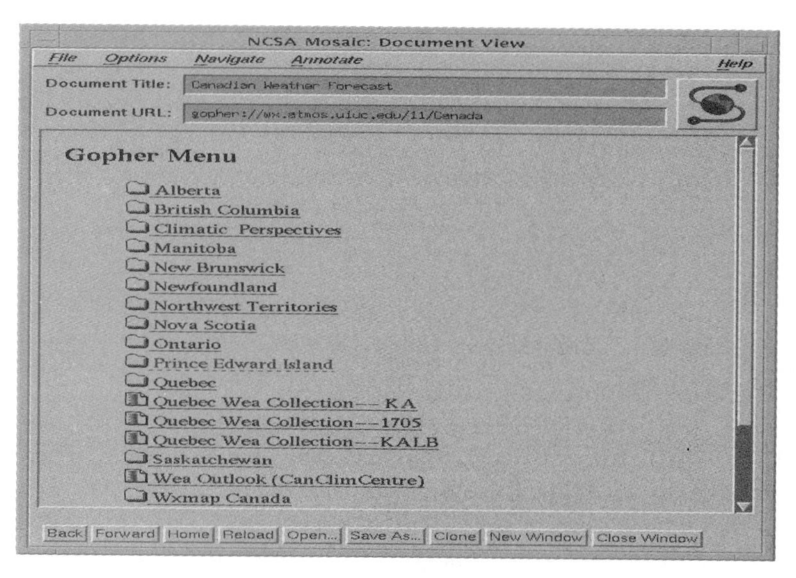

FIGURE 1-7. *A Gopher menu*

The World Wide Web was developed at the European Center for Nuclear Research (CERN) in Geneva, Switzerland. Figure 1-8 shows a screen providing a general overview of the CERN Web. Presently more than 2,300 WWW servers are

NCSA Mosaic: Document View
File Options Navigate Annotate Help
Document Title: Overview of the Web
Document URL: http://info.cern.ch/hypertext/WWW/LineMode/Defaults/defaul

General Overview of the Web

There is no "top" to the World-Wide Web. You can look at it from many points of view. Here are some places to start.

by Subject
The Virtual Library organises information by subject matter.
List of servers
All registered HTTP servers by country
by Service Type
The Web includes data accessible by many other protocols. The lists by access protocol may help if you know what kind of service you are looking for.

If you find a useful starting point for you personally, you can configure your WWW browser to start there by default.

See also: About the W3 project.

Data transfer complete.
Back Forward Home Reload Open... Save As... Clone New Window Close Window

FIGURE 1-8. *The CERN Web*

in operation, connecting both to each other and to other Internet interfaces, including ftp and Gopher. W3 has two unique features that tend to make it quicker and less traumatic to locate the information you need: It catalogs resources by subject, and more importantly, it is based on hypertext.

Hypertext is electronic information linked in a web-like fashion, much like Windows-based help systems. For example, when you are reading document A and come across an item of interest, you can click on it to access a related document B. Instead of following a predetermined order, your information search follows your needs. Perhaps you encounter a word you don't understand. Perhaps you require background information before pursuing the main subject. Perhaps you find a more interesting subject than the original one. The whole point is, you decide. Follow the Web where it leads you.

As an example, you might first access a World Wide Web screen like the one shown in Figure 1-9. From here you could click on CommerceNet to find out more about companies participating in tests of Internet electronic commerce. To find out more, simply keep browsing by clicking on additional topics of interest.

Some Business Applications for the Internet

Subsequent chapters show you how to milk the Internet for data either for your business or for your own entertainment. For starters, let's look at some of the

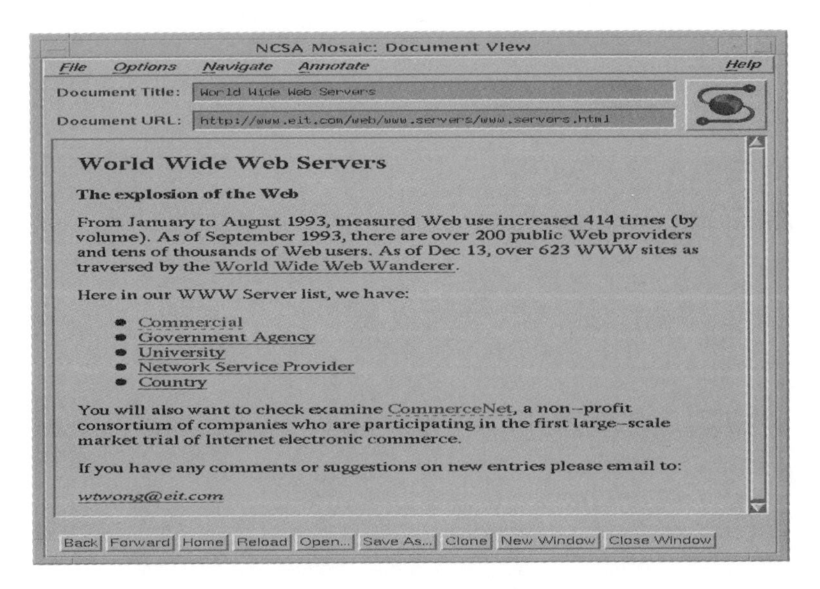

FIGURE 1-9. *Starting a search on the World Wide Web*

specific ways computer manufacturers apply the Internet to improve their competitive position.

Electronic Mail

A major computer company may send literally millions of electronic mail messages per month. The recipients include employees, customers, business partners, and the press. The messages range from a few words, such as a name and telephone number, to entire newsletters.

Product Information

Potential customers can download descriptions of thousands of hardware and software products from the Internet. (*Downloading* simply means transmitting a copy of information from the remote source to a local computer.) The documentation includes technical specifications, brochures, video presentations, technical journals, and catalogs.

Electronic Ordering

Once you're ready to buy, you merely fill out onscreen forms and transmit them electronically. This makes it faster and easier to fill orders, diminishes customer dissatisfaction, and gives salespeople more time to analyze purchase information.

User Support

In the not-so-distant past, computer installations in remote locations were at the mercy of airline schedules for technical help. A minor snowstorm could effectively shut down a major computer system by preventing technical support personnel from arriving on the scene. Via the Internet, technical personnel can test and debug the system, snowstorm or not. If technical support can't solve the problem alone, they can get additional help from senior specialists anywhere in the world. Once the problem is solved, they can send software patches to the remote site electronically. (As of this writing, the Internet still cannot make it stop snowing, however.) Hewlett-Packard (HP) also provides online services and support via the Internet, as shown in Figure 1-10. Other companies such as Microsoft and IBM have similar plans.

What You Need

To access the Internet, you first need to have an Internet account. If you are affiliated with an organization which has an Internet connection, you can use that, or you can obtain your own account from a commercial service known as an Internet provider.

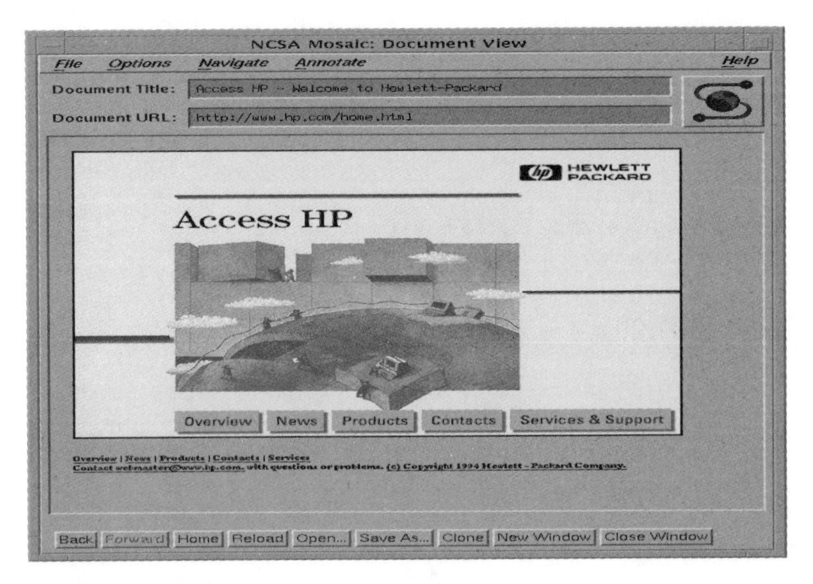

FIGURE 1-10. *Getting help for HP systems*

While it is used by millions, the Internet is not a typical personal computer product. You can't get *full* access to the Internet equipped only with a small microcomputer, a modem, and communications software such as Chameleon. The difficulty does not lie with the microcomputer itself; any model will work (although you'll need a big, fast hard disk if you want to download files). But connecting to the Internet requires the specialized communications protocols described next.

TCP/IP

Computer network communication relies on *protocols,* sets of rules and procedures that enable information exchange. The Internet is based on the *Transmission Control Protocol/Internet Protocol,* commonly abbreviated as *TCP/IP.* This protocol assures that message segments are delivered from one network location to another in the order in which they were sent. It enables dissimilar types of computers and networks to communicate, checks for errors, and initiates retransmission when necessary. TCP/IP doesn't run on a typical microcomputer for the following two reasons:

- TCP was designed to run over cables, either *coaxial cables* similar to those attached to your cable TV or *fiber-optic cables,* hair-thin strands that transmit light signals. In contrast, the modem attached to your microcomputer transmits signals over telephone wires.
- TCP/IP is closely associated with the Unix operating system, which is considerably more sophisticated than DOS or Windows. (Windows NT is a direct competitor to Unix.) Neither DOS nor Windows comes with TCP/IP, although there are TCP/IP versions for both. The usual solutions for making a direct connection between a standard microcomputer and the Internet are described next.

SLIP and PPP

The *Serial Line Internet Protocol (SLIP)* and the *Point-to-Point Protocol (PPP)* can connect microcomputers to the Internet via telephone lines and a modem, preferably a modem that can run at 14400 baud. This provides the computer with its own Internet address, enabling full access to Internet services. (Internet addresses are described in detail in a moment.) The two protocols are similar and are available in public domain and commercial versions. SLIP is older and less sophisticated, while PPP incurs greater transmission overhead.

Internet Addresses

Each computer in a TCP/IP network is identified by a unique 32-bit address, known as an *IP address.* The IP address is a number whose four components are separated by dots—for example, **102.74.115.96**. The IP address consists of three parts: the network ID, the host ID, and the optional subnet ID for very large networks. The *network ID* is assigned by the Network Information Center (NIC), a network clearinghouse. The other addresses are assigned by individual network management.

Networks associated with the Internet use two types of names. The *domain name* identifies your *domain,* a group of machines administered as a single entity. The domain name must conform to the Internet naming conventions and be registered with the Network Information Center (NIC). The *system name,* also called *network node name,* is assigned internally. Table 1-1 presents the top-level domain names associated with the Internet. The system name is used to set up addresses within the internal network. It is required by TCP/IP services such as e-mail, ftp, and telnet.

The Internet is a worldwide phenomenon. Table 1-2 lists some of the major international domain names.

DOMAIN NAME	DESCRIPTION	EXAMPLE
com	Commercial organizations	**sun.com** for Sun Microsystems
edu	Educational institutions	**cmu.edu** for Carnegie-Mellon University
gov	Government organizations	**locis.loc.gov** for Library of Congress
mil	Military organizations	**navnews@nctamslant.navy.mil** for Navy News Service
net	Network providers and information centers	**commerce.net** for CommerceNet
org	Nonprofit organizations	**bbhost.hq.eso.org** for European Southern Observatory Bulletin Board

TABLE 1-1. *Internet Top-Level Organizational Domain Names*

DOMAIN NAME	COUNTRY
au	Australia
at	Austria
ca	Canada
cl	Chile
dk	Denmark
fr	France
de	Germany
il	Israel
it	Italy
jp	Japan
kr	South Korea
es	Spain
se	Sweden
uk	United Kingdom/Ireland
us	United States

TABLE 1-2. *Internet Geographical Top-Level Domain Names*

Internet Management Issues

Vinton Cerf, the "father of the Internet," foresees a world in which every telephone has a hookup to the Internet. However, he is not just cheerleading. As he puts it, this vision is his "ultimate either dream or nightmare." Perhaps he was thinking of some of the management issues and potential problems discussed next.

Billing

A major Internet problem is that familiar bugaboo: money. Someone must pay for an online search stretching from Ramona in Pamona to Kokomo to Tokyo to LouLou in Honolulu. The question is who, and the answer is not always clear. Computer services that seem free can be expensive. Part of the problem is the vast quantity of available information. Discipline is a necessity, as many subscribers to online services such as CompuServe can readily testify. When calculating the cost of using the Internet, don't forget to count the cost of training, and the employee time and connection costs incurred when someone is wandering through the Internet, hoping to find something. This activity is known as *surfing,* and it can be expensive. But no matter how services are charged, it is considered good online etiquette not to waste or tie up resources unnecessarily.

Security

While frustrated surfers may disagree, the whole point of the Internet is easy access to information and programs. Ease of access raises the possibility of security breaches, whether malicious, playful, or unintentional. In 1988 a skillful hacker set off the famous Internet worm, crashing thousands of computer networks. Some people feel that, as with earthquakes in California, the "big one" is still to come.

There is no simple formula for eliminating the Internet security risk short of permanently disconnecting from it, an option that nobody recommends. The following guidelines can increase Internet security as well as computer network security in general.

- **Passwords:** A *password* is a secret code by which a user identifies himself or herself to the system. Many people consider properly controlled passwords to be the single most important way of increasing system security. To be effective, passwords must be carefully chosen (no nicknames or pet names please), guarded, confidential, and changed on a regular basis.

- **Recording unsuccessful logons:** The system administrator responsible for the Internet should record unsuccessful attempts at system access. One or two such attempts may not be a problem. But an unauthorized individual consistently trying to access the Internet, or off-limit areas on the Internet, should be monitored.
- **Isolating the Internet computer:** Although this can cause a bottleneck, many organizations funnel all Internet access through a single computer. This computer serves as a *firewall*, protecting other computers from intrusion much as a physical firewall protects people and property from a raging fire. Of course, the firewall (computer) must be carefully monitored.
- **Data Encryption:** *Data encryption* involves coding data so that unauthorized individuals cannot read it, increasing security and privacy. This technique has important military and commercial consequences. Data encryption software is available for all budgets.
- **Commitment:** The preceding techniques for maintaining security, and any other techniques not mentioned here, are worthless unless everyone who uses the computer makes a commitment to computer security.

Access

A company or workgroup should define and implement Internet access guidelines for its employees. Otherwise, they may waste valuable resources on frivolous activities such as **alt.barney.die.die.die**, an Internet discussion group devoted to the destruction of that ever-popular dinosaur, Barney. While we will refrain from supplying any references, the Internet provides many more controversial discussion groups.

However, excessively restricting employee Internet access can be like throwing out the baby with the bath water. Who can be sure where to find useful documents or programs? Nothing can be more frustrating than tracking down an item across sources and continents, only to be informed that your access has been denied. Furthermore, people learn by doing. The navigation principles and shortcuts acquired in accessing World Cup '94 information later can be applied to accessing legal and technical documentation for storing liquid nitrogen in Rio de Janeiro, as just one example.

Privacy

In some ways the Internet is the antithesis of privacy. To a certain degree, appropriate security and access controls can stop people from reading documents that they have no business reading. At the same time, if you don't want something to be read by just anybody, you probably shouldn't put it on the Internet. It may help to think of your e-mail and any other documents or communication on the Internet as postcards that anyone can read, rather than as letters directed to one specific individual.

An interesting issue is whether an organization should be able to read its employees' Internet messages. Some oppose snooping, pointing out that secretly reading employees' mail will be counterproductive. Once employees know their messages can be read by unintended parties, they won't express their true feelings by mail, effectively destroying the system's value. The danger is not far-fetched; employees have been fired after their electronic mail was read.

Others insist that the employer owns the network and therefore has the right to monitor the messages as required. Given the power of the Internet, unscrupulous employees could damage the firm, perhaps irreparably, within a matter of minutes. Management must protect the firm; if not, stockholders may sue. Employees who want to send private messages should do so on their own time, using their own Internet connection. As you can see, the matter is up for debate.

Introducing Mosaic

You can use a single Mosaic interface to navigate across the Internet and view selected documents. This interface employs hypertext and hypermedia. As you learned earlier, *hypertext* is electronic text linked in a web-like fashion. *Hypermedia* extends the linked information to include multimedia features such as graphics and audio. You can navigate from document to document by clicking on electronic links called *hyperlinks.* The actual physical process of accessing documents and copying (downloading) their contents is transparent. This allows you to focus on the documents themselves, rather than on the search process.

Mosaic is based on the *client-server architecture.* This architecture divides the processing of an application between a client, which processes data locally and maintains a user interface, and a *server,* which handles database and computing-intensive processing.

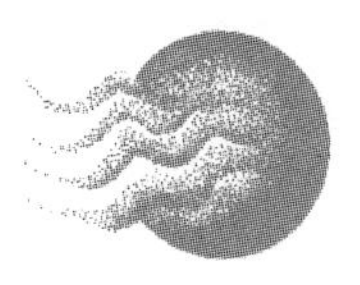

NOTE
Multiple clients can connect to the same server.

The server is an Internet information source. It responds to *queries* (information requests) issued by Mosaic clients located anywhere on the Internet. Mosaic transparently integrates information from a wide variety of Internet servers, including WWW servers, Gopher, and anonymous FTP (File Transfer Protocol) servers. You needn't know that a particular query accessed different types of servers in geographically distinct locations; Mosaic hides these ugly details.

As you gain confidence and experience, you can upgrade Mosaic with additional viewers, enabling Mosaic to process sophisticated data types such as movie files and Postscript files. Appendix B demonstrates how to add a viewer to Mosaic to increase its capabilities. Mosaic evolves with the changing information environment, and is easy to customize to meet the specific information retrieval needs of individuals, workgroups, and organizations.

CHAPTER 2

Getting Started with the Windows Version of Mosaic

This chapter is for those who want to run Mosaic on Microsoft Windows systems—either Windows 3-1 or Windows NT. After providing a general introduction to Mosaic, the chapter describes what you need to run Mosaic. After that is a quick but thorough overview of the Mosaic menus. When you finish this chapter, you'll be ready to go on a tour of Mosaic, which starts in Chapter 4.

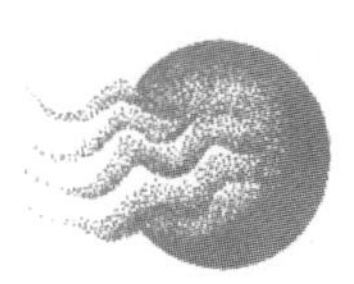

NOTE
If you are using Mosaic in X Window, you can skip this chapter and go on to Chapter 3.

Installing and Using Mosaic

Mosaic runs on the three major types of desktop computers: MS-Windows (IBM-compatible microcomputers), Macintosh, and X Window for Unix-based workstations. As you might expect, the procedures for installing and running Mosaic differ somewhat for these three platforms. This section describes what it takes to get Mosaic for MS-Windows up and running. Appendix B presents detailed installation procedures for the MS-Windows and X Window versions.

The developers of Mosaic, the National Center for Supercomputing Applications (NCSA), have placed a free, copyrighted version of Mosaic on the Internet that you can download for individual use. NCSA's address for anonymous ftp transfers is **ftp.ncsa.uiuc.edu**. Several companies have licensed Mosaic and are developing enhanced, fully supported versions for sale. Spyglass, Inc. intends to relicense Mosaic in high-volume quantities to many vendors, such as the Digital Equipment Corporation (DEC), who will incorporate it into their products.

Connecting to the Internet

You cannot run Mosaic with a slow modem. Furthermore, Mosaic requires a computer with direct TCP/IP Internet access, PPP (Point to Point Protocol), or a SLIP (Serial Line Internet Protocol) connection.

TIP
Many Internet applications involve a lot of overhead. The faster the modem the better. Use a 14,400 baud modem or, at a minimum, a 9,600 baud modem. Don't use a 2,400 baud modem; it's too slow.

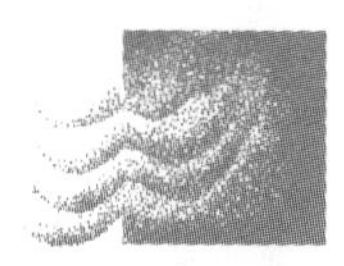

CAUTION
Your connection is only as fast as the modem/computer on the other end.

Direct TCP/IP Internet access means that the computer is connected to the Internet via a leased or dedicated line running the TCP/IP communications protocol or is connected to a network that is connected in this manner.

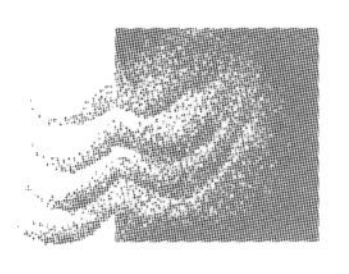

CAUTION

If your computer dials into the Internet provider, make sure that the account provides full Internet connectivity via PPP or SLIP.

Testing NCSA Mosaic

Software installation is only the tip of the iceberg. What's important is testing it. If you have trouble, you can get help from NCSA, provided that you have correctly specified your e-mail address to receive responses.

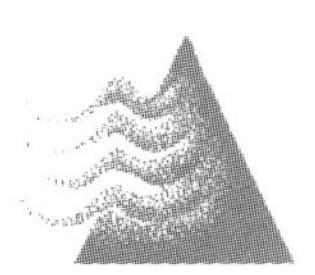

TIP

If you really want extensive support, you must purchase a commercial version of Mosaic.

Test the Mosaic installation by double-clicking on the Mosaic icon. This should generate a screen similar to the one shown in Figure 2-1. In this case, you have accessed the default NCSA Mosaic for Microsoft Windows *Home Page,* or startup

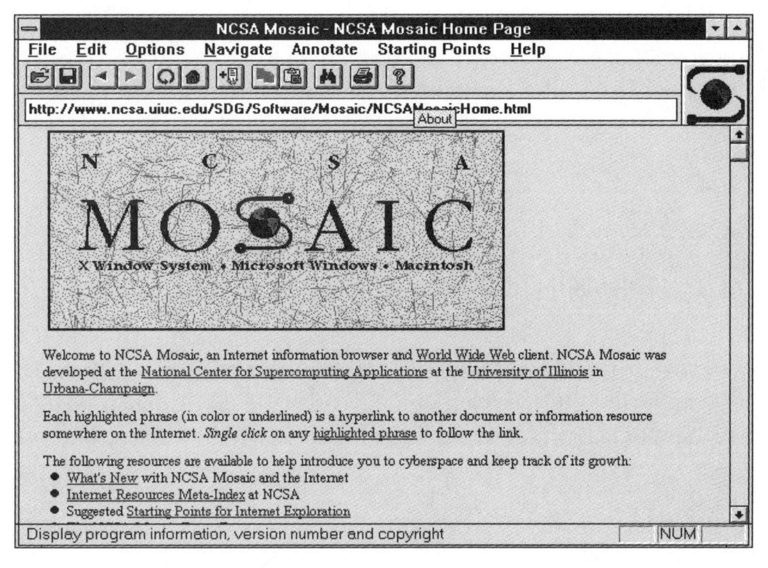

FIGURE 2-1. *NCSA Home Page*

document. See Appendix B for a discussion of the common installation errors and how to correct them.

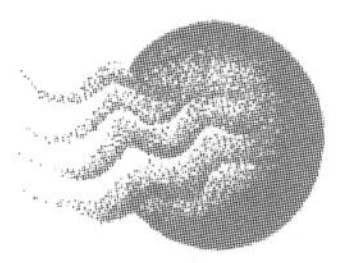

NOTE
Chapter 5 explains how to create and modify your own Home Page, one of the most important ways to customize Mosaic to meet your needs and, as luck would have it, increase your productivity at the same time.

The Mosaic Main Menu

As might be expected, the look and feel of Mosaic for Windows menus and Mosaic for Unix menus will be familiar to users of Windows and Unix. The two sets of menu options are similar, but do have important differences. Learn one version of Mosaic and you will have no trouble making the transition to the other version.

Mosaic for Windows Main Menu

The Mosaic for MS-Windows main menu is shown here:

File Edit Options Navigate Annotate Starting Points Help

The menu bar contains seven menu items, ranging from File to Help. Selecting a menu item generates a pull-down menu. You can pull down most menus by using either the mouse or the keyboard. In all cases you can open a menu by clicking on the menu item.

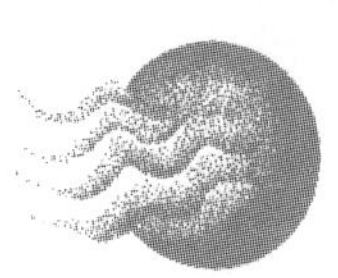

NOTE
Your Mosaic main menu may include more or fewer menu items, depending on how it was set up.

In most cases, you can hold down the ALT key and press the underlined letter in the option name to pull down a menu. For example, to display the File menu, shown here, either click on the word *File* or hold down the ALT key and press F.

File
Open URL... Ctrl+O
Open Local File...
Save Ctrl+S
Save As...
Save Preferences
Print... Ctrl+P
Print Preview
Print Setup...
Document Source...
Exit

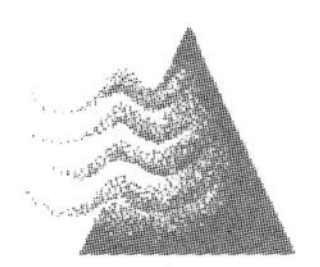

TIP
When you press a letter in combination with the ALT, SHIFT, or CTRL key, it doesn't matter whether the letter is uppercase or lowercase.

Windows applies certain conventions to its pull-down menus. Menu items that are *dimmed* (or *grayed*) are presently unavailable. For example, initially the Save option is dimmed; you cannot save a file if you haven't opened one yet. Menu items followed by an ellipsis (...), such as Print, lead to a dialog box requesting additional information. Menu items with a triangle to their right lead to additional menus. You can issue some commands, such as Print, directly from the keyboard (without going through the menus)—in this case by holding down the CTRL key and pressing P. The horizontal lines in the menu group together related menu selections, such as the various print options. Next you learn about the seven pull-down menus and the various options that they provide.

The File Menu

You use the File menu to load documents directly from the Internet or from local disks—for example, when temporarily storing search results. The File menu includes ten options, which are described next.

- **Open URL...** opens a Uniform Resource Locator (URL), a standardized address used to locate information anywhere on the Internet. Clicking on this option generates the dialog box shown here:

Open URL
URL:
Current Hotlist: OK Cancel

Chapter 5 describes URLs in greater detail.

- **Open Local File...** enables you to open a local document, one that you can access directly.
- **Save** saves the current document, the one on which you are presently working.
- **Save As...** saves the current document with a file name that you specify and confirm.
- **Save Preferences** saves choices you made when customizing the Mosaic interface.
- **Print...** prints a specified document.
- **Print Preview** shows you what your printed document will look like.

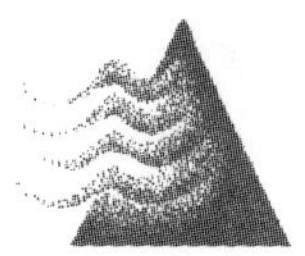

TIP
Judicious use of the Print Preview option can save you time and paper.

- **Print Setup...** selects the printer and specifies print information such as paper size.
- **Document Source...** gives you an internal view of your Mosaic document. It displays the names and addresses of referenced documents instead of their hyperlinks. Spend a little time comparing Figure 2-1 to its internal view, which is shown in Figure 2-2. Underlined phrases in Figure 2-1 appear in Figure 2-2 with a URL to their left. Clicking on World Wide Web references the **http://info.cern.ch/hypertext/WWW/TheProject.html** file. Such a request links to a file in Geneva, Switzerland, the home of CERN.
- **Exit** lets you leave Mosaic.

The Edit Menu

You use the Edit menu, shown here, to copy or find a document segment such as a *character string,* a series of characters. The Edit menu offers two selections, which are described next.

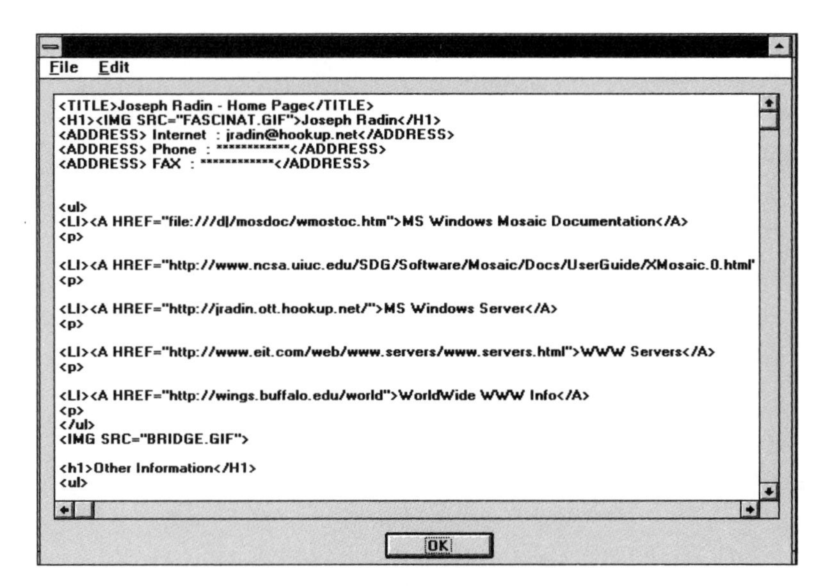

FIGURE 2-2. *Internal view of the NCSA Home Page*

- **Copy** copies a selected character string to the Clipboard, from which it can be copied to the same document or other documents.
- **Find...** locates the next occurrence of the specified character string in the designated document. Selecting this option displays the Find dialog box shown here:

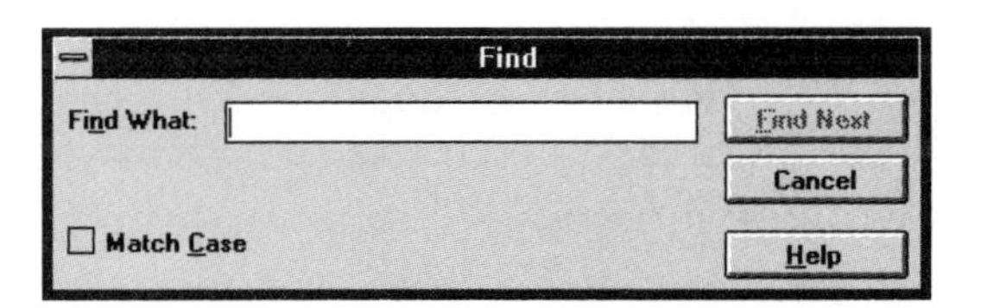

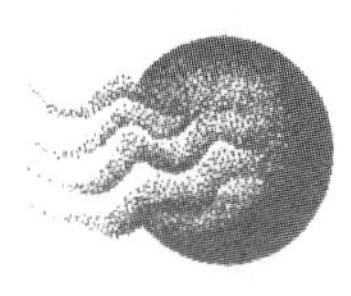

NOTE
You can indicate whether you want the search to be case sensitive.

The Options Menu

You use the Options menu, shown here, primarily to customize the screen's appearance. Many of these menu items are *toggles:* selections with the two states on (as indicated by a check mark) and off. Each time you click on a toggle, you change its state. The Options menu offers 12 selections, most of which are described in greater detail in later chapters.

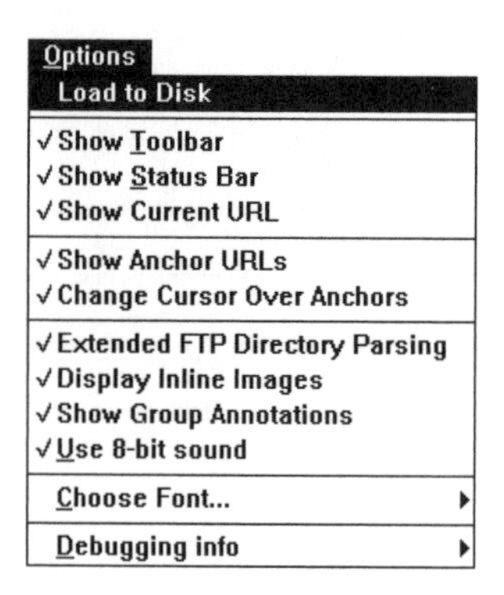

- **Load to Disk** enables you to specify that you want to save documents to a (local) disk rather than displaying them.
- **Show Toolbar** shows or suppresses the toolbar, a handy bar near the top of the screen that contains small images called *icons*. These icons are mouse shortcuts for frequently executed commands. For instance, you can click on the file folder icon to open a file. You can see the toolbar on the third line in Figure 2-1.
- **Show Status Bar** shows or suppresses the status bar, a line at the bottom of the screen that displays system information such as the number of bytes transferred. The status bar also indicates when the CAPS LOCK, NUM LOCK, or SCROLL LOCK key is on.
- **Show Current URL/Title** shows or suppresses the document title and its Uniform Resource Locator. You should make sure that this option is on. This information is useful if you get lost.
- **Show Anchor URLs** shows or suppresses the underlining or special color for hyperlinks such as World Wide Web. It's a good policy to keep this toggle switch on. Hyperlinks are very important and should be readily visible.
- **Change Cursor Over Anchors** shows or suppresses a different cursor shape when the cursor is over an underlined term. You should keep this toggle switch on since it's important that hyperlinks be readily visible.
- **Extended FTP Directory Parsing** is a technical productivity consideration. It controls the display of file and directory names.

- **Display Inline Images** shows or suppresses images in a transmitted document. Suppressing the image—that is, turning this option off—reduces the transmission time.
- **Show Group Annotations** shows or suppresses group annotations (comments made by other members of your workgroup). See the later section on the Annotate menu for additional details.
- **Use 8-Bit Sound** uses or suppresses sound. Given Mosaic's commitment to keeping up with technology, future versions should offer more sophisticated audio capabilities.
- **Choose Font...** leads to the submenu shown in Figure 2-3. Enables you to select among the font types defined in the **mosaic.ini** file, the Mosaic customization file which is discussed in detail in Chapter 7.
- **Debugging Info** leads to the submenu shown in Figure 2-4. This menu permits specialists to obtain more information when problems occur.

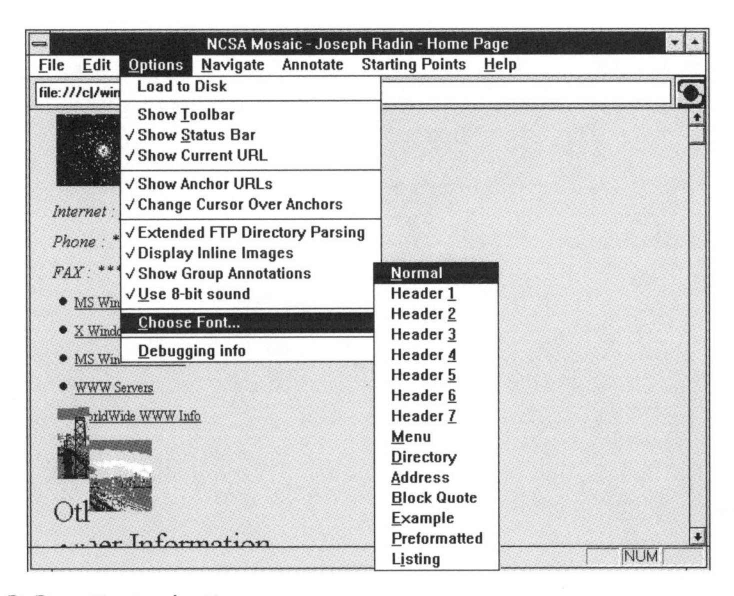

FIGURE 2-3. *Font selection*

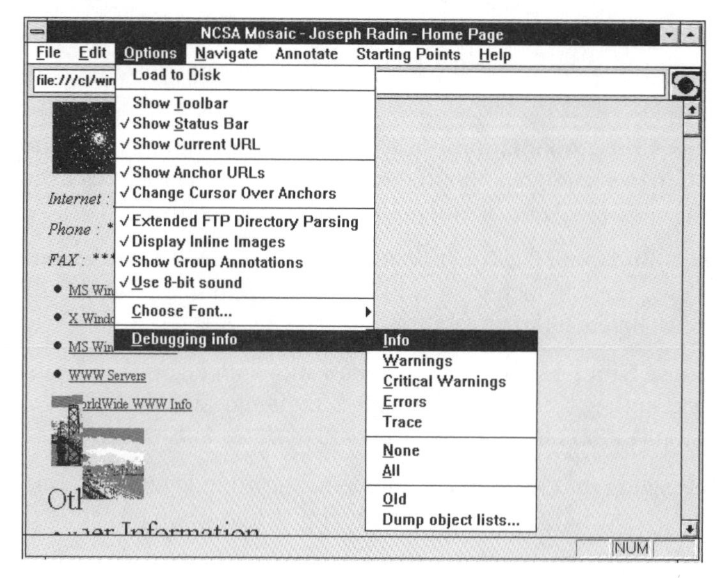

FIGURE 2-4. *Debugging info menu*

The Navigate Menu

The Navigate menu, shown here, lets you maneuver efficiently across the Internet. The Navigate menu offers seven options, as described next.

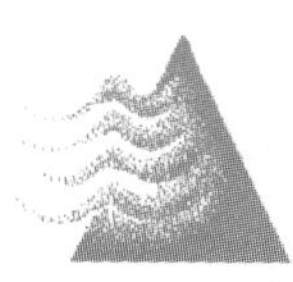

TIP

The time and energy you invest in mastering the Navigate menu will be paid back in spades.

- **Back** retrieves the most recently accessed document. Depending on how many documents you've retrieved in the current work session, you may be able to click on this item several times.

- **Forward** is the opposite of Back. Use this menu item when you press Back too many times and overshoot the desired document.
- **Reload** reloads the current document.
- **Home** returns you to your Home Page.
- **History** displays a History window like the one shown in Figure 2-5; this window includes a list of the URLs for documents in the order in which they were accessed. Select any item in the list and press the Load button to obtain it. Press the Dismiss button to close this window.
- **Add Current to Hotlist...** places the document that you are working on in the Hotlist that contains documents you retrieve often.
- **Menu Editor...** opens a window for editing menus as discussed in Chapter 8.

The Annotate Menu

You use the Annotate menu, shown here, to add your own comments to Mosaic documents for future reference. The Annotate menu offers the three selections discussed next.

Annotate
Annotate
Edit this annotation
Delete this annotation

- **Annotate** generates the special Annotate Window shown in Figure 2-6; here is where you compose your comments for the given file. You can enter *personal comments,* in which case only you can read them. You can also enter *group comments,* which anyone in your workgroup can read provided that the Show Group Annotations item on the Options menu is on (has a checkmark to its left). Finally, you can enter *public comments,* which anyone with access to the given document can read. This menu option is discussed in greater detail in Chapter 8.
- **Edit This Annotation** enables you to change the contents of the designated annotation.
- **Delete This Annotation** removes the designated annotation, permanently.

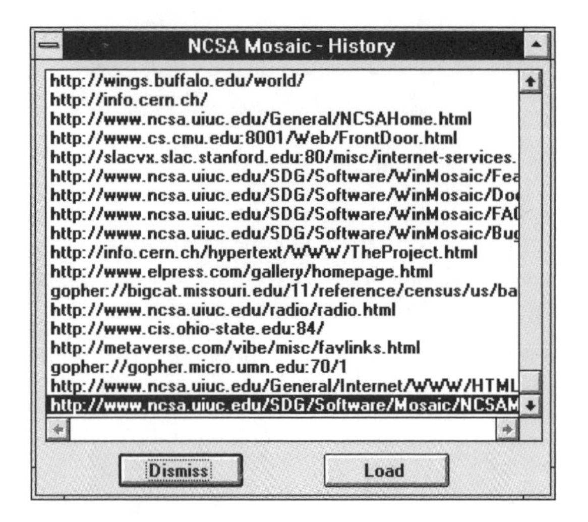

FIGURE 2-5. *The History option*

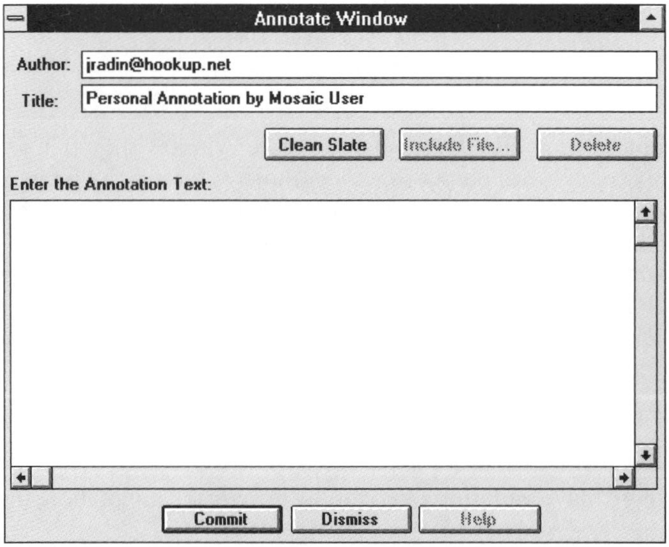

FIGURE 2-6. *Composing a personal annotation*

The Starting Points Menu

The Starting Points menu, shown here, lets you specify where to begin a search, saving you time. This menu is used extensively in Chapter 5. Chapters 7 and 8 discuss in detail how to set up such a menu.

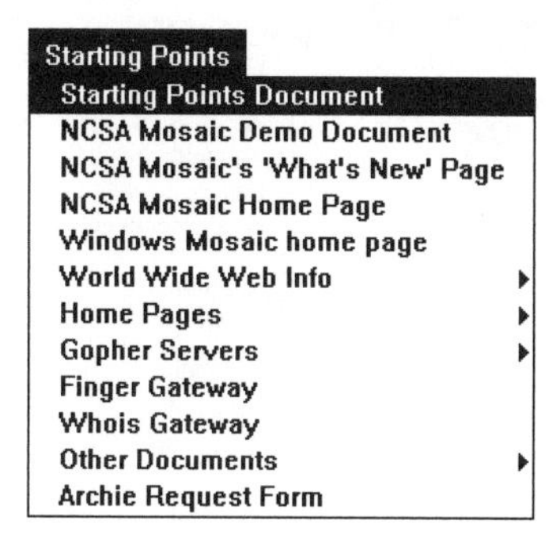

Menu items include:

- **World Wide Web Info** Figure 2-7 shows the World Wide Web Info submenu.
- **Gopher Servers** Figure 2-8 shows the Gopher Servers submenu.
- **Other Documents** Figure 2-9 shows the Other Documents submenu.

The Help Menu

You use the Help menu, shown here, to obtain help when you need it. One reason for Mosaic's explosive growth is that its Help screens are so useful and so easy to find. The Help menu offers these six selections:

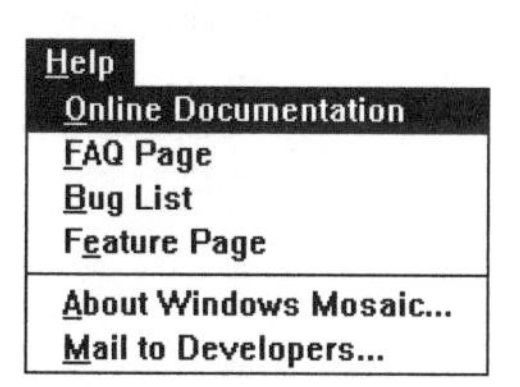

- **OnLine Documentation** provides explanations at the click of a mouse.
- **FAQ Page** answers frequently asked questions about Mosaic.
- **Bug List** lists known Mosaic errors.

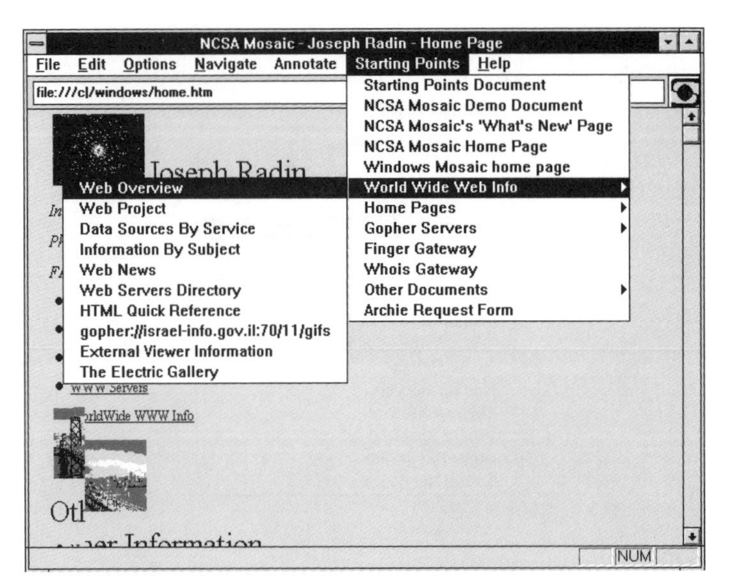

FIGURE 2-7. *The World Wide Web Info submenu*

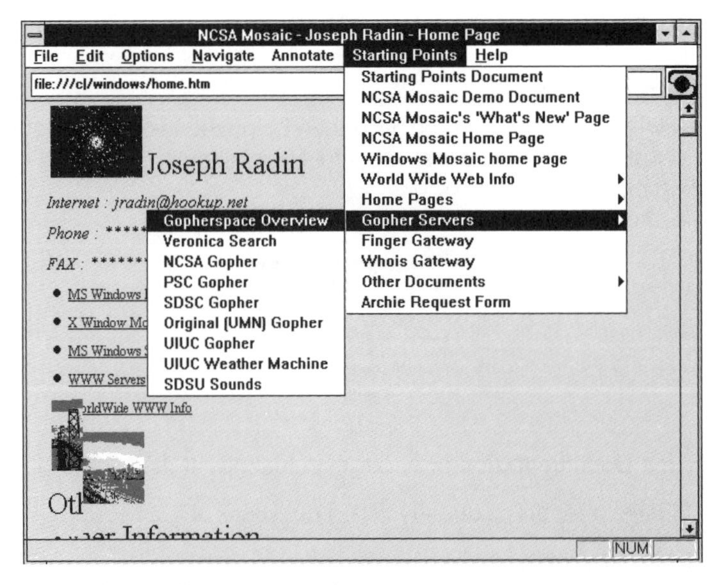

FIGURE 2-8. *The Gopher Servers submenu*

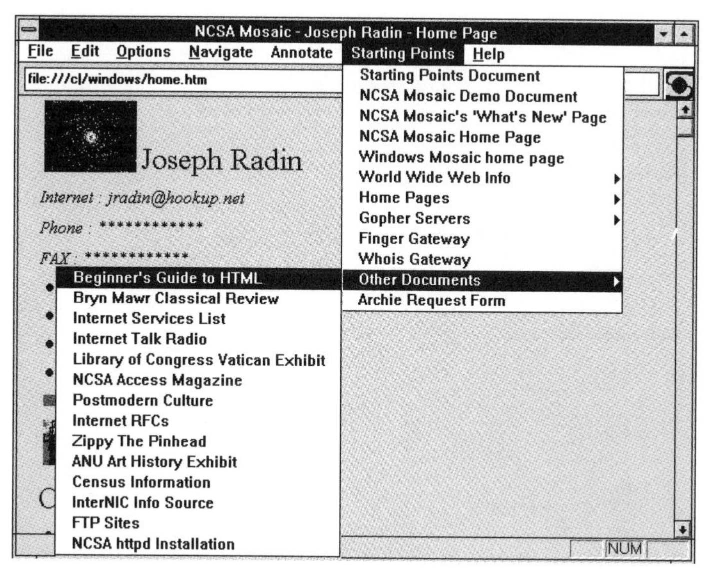

FIGURE 2-9. *The Other Documents submenu*

- **Feature Page** contains general information about Mosaic features.
- **About Windows Mosaic...** provides basic information about this Windows Mosaic product.
- **Mail to Developers...** helps you contact the NCSA technical specialists when you need help by sending them a mail message.

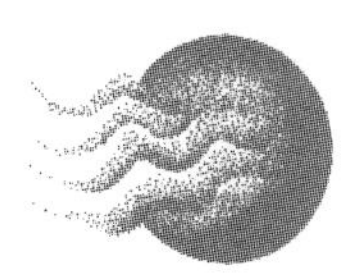

NOTE
Since the NCSA version of Mosaic is public domain software with limited user support, don't expect an immediate response from a technical specialist. We hope you'll find that this book meets your needs. You should also consider acquiring a supported version of Mosaic.

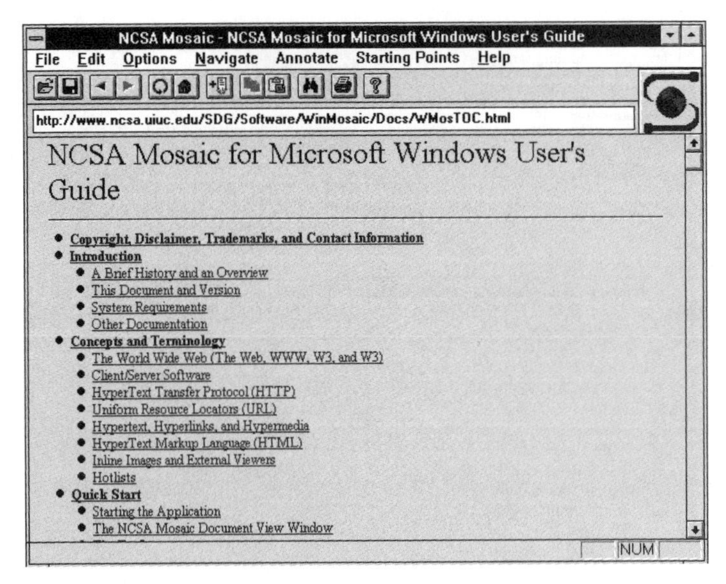

FIGURE 2-10. *The User's Guide*

Figure 2-10 shows the User's Guide obtained by clicking on the Online Documentation item in the Help menu. The toolbar appears at the third line of the screen. The toolbar icons are described in Table 2-1.

Now that you've had an introduction to using Mosaic in Windows, you're ready to skip to Chapter 4 for a guided tour.

ICON	DESCRIPTION
Open URL	Opens the URL dialog box and provides access to the Hotlist.
Save	Saves the current document to disk.
Back	Displays the previous document in the history list (unless the current document is the first document).
Forward	Displays the next document in the history list (unless the current document is the last document).
Reload	Reloads the current document.
Home	Returns to the default Home Page.
Add to Hotlist	Adds the current document to the Hotlist.
Copy	Copies the current selection to the Clipboard.
Paste	Pastes the current selection to the Clipboard.
Find	Opens the Find dialog box to find a character string in the current document.
Print	Prints the current document.
About	Opens the About NCSA Mosaic for Microsoft Windows window.

TABLE 2-1. *The toolbar icons*

Getting Started with X Window System (Unix) Mosaic

This chapter is for those who want to run Mosaic on X Window systems (which are associated with the Unix operating system). After providing a general introduction to Mosaic, the chapter tells you what you need to run Mosaic. Next is a quick but thorough overview of the Mosaic menus. When you finish this chapter, you'll be ready to take a tour of Mosaic, which starts in Chapter 4.

Installing and Using Mosaic

Mosaic runs on the three major types of desktop computers: MS-Windows (IBM-compatible microcomputers), Macintosh, and X Window for Unix-based workstations. As you might expect, the procedures for installing and running Mosaic differ somewhat for these three platforms. This section describes what it takes to get Mosaic for MS-Windows up and running. Appendix B presents detailed installation procedures for the MS-Windows and X Window versions.

The developers of Mosaic, the National Center for Supercomputing Applications (NCSA), have placed a free, copyrighted version of Mosaic on the Internet that you can download for your own use. NCSA's address for anonymous ftp transfers is **ftp.ncsa.uiuc.edu**. Several companies have licensed Mosaic and are developing enhanced, fully supported versions for sale. Spyglass, Inc. intends to relicense Mosaic in high-volume quantities to many vendors, such as the Digital Equipment Corporation (DEC), who will incorporate it into their products.

Connecting to the Internet

Mosaic requires a computer with direct TCP/IP Internet access, a PPP (Point to Point Protocol), or a SLIP (Serial Line Internet Protocol) connection.

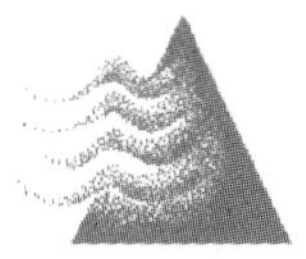

TIP
Use a 14,400 baud modem or, at a minimum, a 9,600 baud modem. Don't use a 2,400 baud modem; it's too slow.

Direct TCP/IP Internet access means that the computer is connected to the Internet via a leased or dedicated line running the TCP/IP communications protocol or is connected to a network that is connected in this manner.

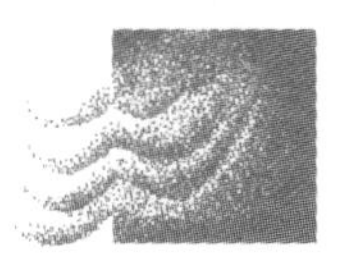

CAUTION
If your computer dials into the Internet provider, make sure that the account provides full Internet connectivity via PPP or SLIP.

Testing NCSA Mosaic

Software installation is only the tip of the iceberg. What's important is testing it. If you have trouble, you can get help from NCSA, provided that you have correctly specified your e-mail address to receive responses.

Start Mosaic by entering the following at the Unix prompt:

```
xmosaic
```

You should see a screen similar to the one shown in Figure 3-1. In this case, you have accessed the default NCSA Mosaic for X Window Home Page, or startup document.

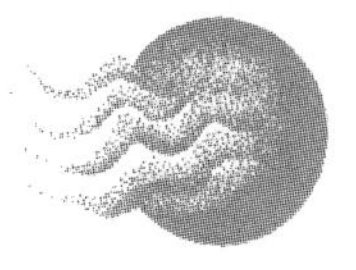

NOTE
Chapter 5 demonstrates how to create and modify your own Home Page, one of the most important ways to customize Mosaic to meet your needs and, as luck would have it, increase your productivity at the same time.

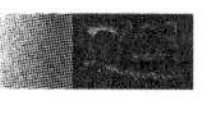

The Mosaic Main Menu

As you might expect, the look and feel of Mosaic for Unix menus and Mosaic for MS-Windows menus will be familiar to users of Unix and MS-Windows. The two sets of menu options are similar, but do have important differences. Learn one version of Mosaic and you will have no trouble making the transition to the other version.

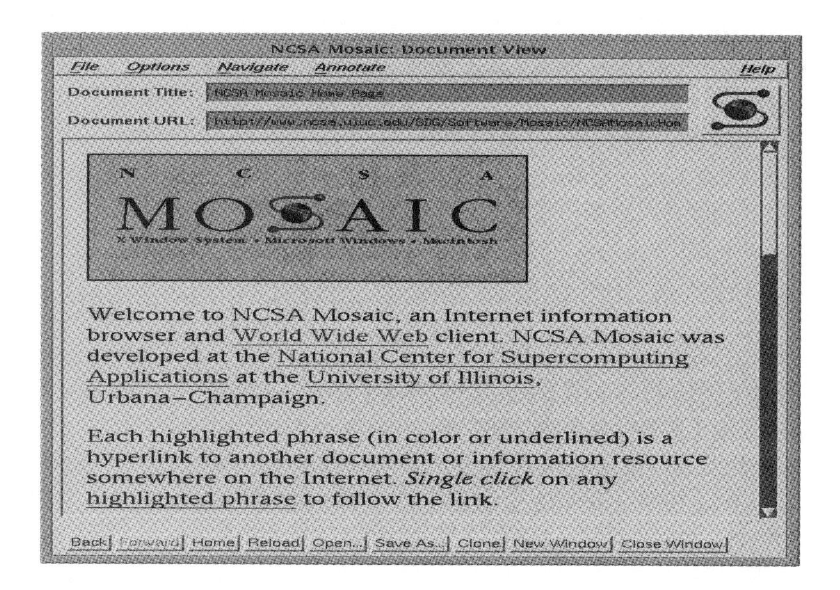

FIGURE 3-1. *NCSA Mosaic for X Window Home Page*

Mosaic for Unix Main Menu

The Mosaic main menu is shown in Figure 3-1. The second line of this menu contains five pull-down menus: File, Options, Navigate, Annotate, and Help (at the far right). In a moment, we will examine the five available menus. Selecting a menu item generates a pull-down menu. You can display most menus by using either the mouse or the keyboard. In all cases you can click on the menu identifier to display the menu in question. In addition, in most cases you can display a menu by holding down the ALT key and pressing the underlined letter in the option name. For example, to generate the File menu, either click on the word *File* or hold down the ALT key and press the letter F.

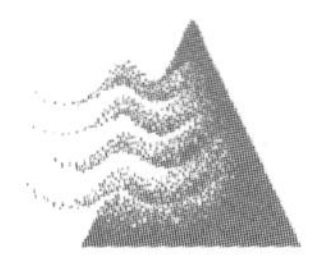

TIP
When you press a letter in combination with the ALT, SHIFT, or CTRL key, it doesn't matter whether the letter is uppercase or lowercase.

Menu items that are *dimmed* (or *grayed*) are presently unavailable. Menu items followed by an ellipsis (...), such as Find or Print, lead to a dialog box requesting additional information. Menu items with a triangle to their right lead to additional menus.

The File Menu

You use the File menu, shown here, to load documents directly from the Internet or from local disks—for example, when temporarily storing search results. The File menu includes 14 options, which are described next.

File
New Window
Clone Window
Open URL...
Open Local...
Reload Current
Reload Images
Refresh Current
Find In Current...
View Source...
Save As...
Print...
Mail To...
Open DTM Outport...
Broadcast Over DTM
Close Window
Exit Program...

- **New Window** opens a new window.
- **Clone Window** opens a copy of an existing window.

- **Open URL...** opens a Uniform Resource Locator (URL), a standardized address used to locate information anywhere on the Internet. Clicking on this option generates the dialog box shown here:

The last line of the window includes four buttons: the Open button opens the designated file, the Clear button erases the window contents, the Dismiss button closes the window without opening a URL, and the Help button generates context-sensitive help. (Several of these buttons are found in other Mosaic pull-down menus as well.) Chapter 5 covers URLs in greater detail.

- **Open Local...** enables you to open a local document, one that you can access directly. A sample dialog box is shown in Figure 3-2. Note the list boxes for selecting the directory and file names.

FIGURE 3-2. *Open Local Document dialog box*

- **Reload Current** gets a copy of the current document if the Home Page has been changed.
- **Reload Images** gets a copy of graphics files linked to the current document if the Home Page has been changed.
- **Refresh Current** redraws the screen for the current document if the Home Page has been changed.
- **Find...** locates the next occurrence of the specified character string in the designated document, as shown here:

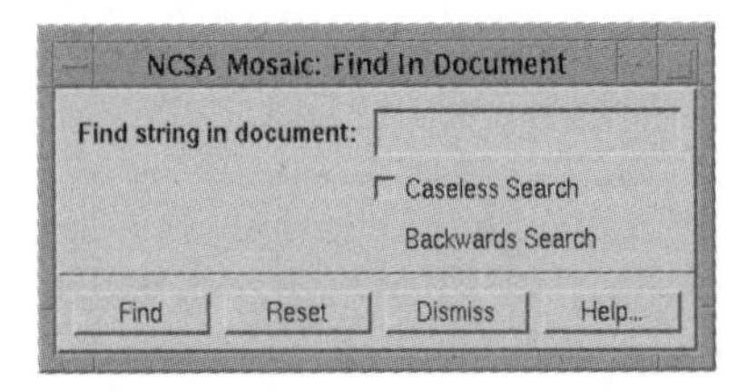

  Click on the Caseless Search box to ignore whether specified characters are upper- or lowercase.

- **View Source...** gives you an internal view of your document. It displays names and addresses of reference documents instead of their hyperlinks. Spend a little time comparing Figure 3-1 and its internal view, which is shown in Figure 3-3. Underlined phrases in Figure 3-1 appear in Figure 3-3 with a URL to their left.
- **Save As...** saves the current document with a file name that you specify and confirm. Figure 3-4 shows an example of this dialog box. You can save the file in one of these four formats: Plain Text, Formatted Text, Postscript, and HTML (Mosaic hypertext format).
- **Print...**, as shown here, prints a specified document in one of these four formats: Plain Text, Formatted Text, Postscript, and HTML (Mosaic hypertext format).

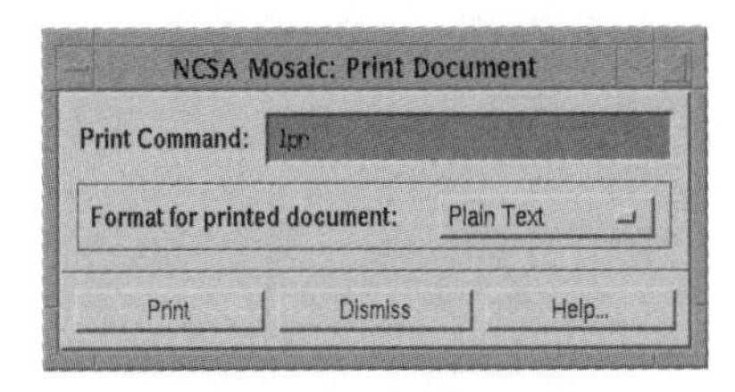

- **Mail To...**, as shown here, lets you send mail.

NCSA Mosaic: Document Source
Uniform Resource Locator: http://www.ncsa.uiuc.edu/SDG/Software/Mosaic/NCSAMosai
<TITLE>NCSA Mosaic Home Page</TITLE>
<IMG SRC="mosaic.gif"> <P>
Welcome to NCSA Mosaic, an Internet information browser and <A
HREF="http://info.cern.ch/hypertext/WWW/TheProject.html">World Wide
Web</A> client.
NCSA Mosaic was developed at the
<A HREF="http://www.ncsa.uiuc.edu/General/NCSAHome.html">National Center
for Supercomputing Applications</A> at the
<A HREF="http://www.ncsa.uiuc.edu/General/UIUC/UIUCIntro/UofI_intro.html">Univer
of Illinois</A>, Urbana-Champaign.
<P>
Each highlighted phrase (in color or underlined) is a hyperlink to
Dismiss
Help...

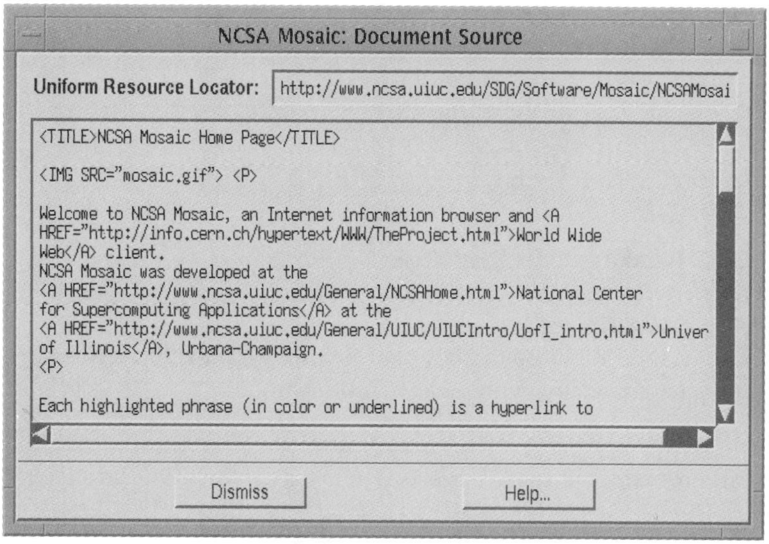

FIGURE 3-3. *Document source*

NCSA Mosaic: Save Document
Filter
/usr/users/joe/*
Directories
..
.mosaic-personal-annotations
DAYTONA
OMH
PROXY
bin
Files
.Xauthority
.Xdefaults
.Xdefaults.old
.cshrc
.login
.merlinserverlist
.mosaic-global-history
.mosaic-hotlist-default
Format for saved document:
Plain Text
Name for saved document:
/usr/users/joe/
OK
Filter
Cancel

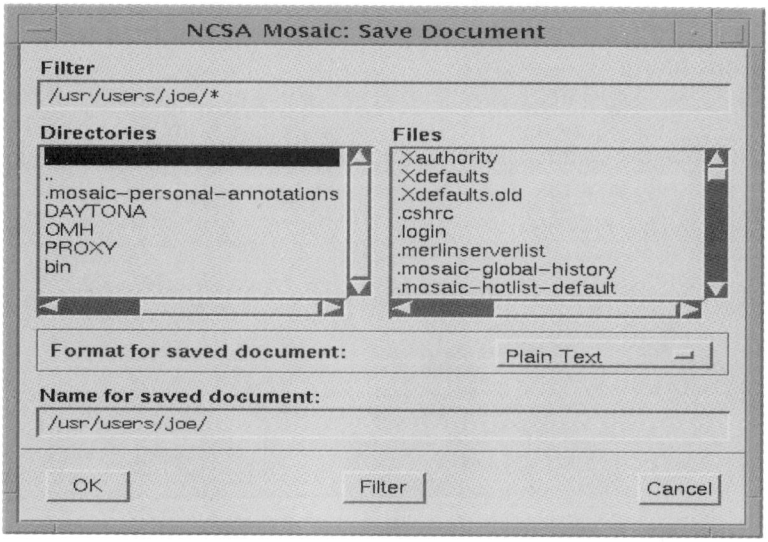

FIGURE 3-4. *Save Document dialog box*

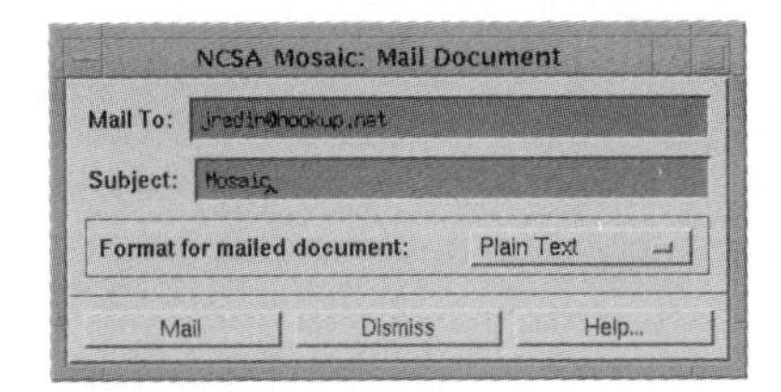

- **Close Window** closes a window.

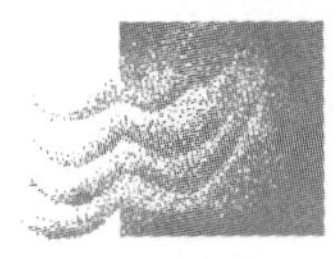

CAUTION
When you close the last open window you exit from Mosaic automatically, without any confirmation.

- **Exit Program...** lets you leave Mosaic after confirmation, as shown here.

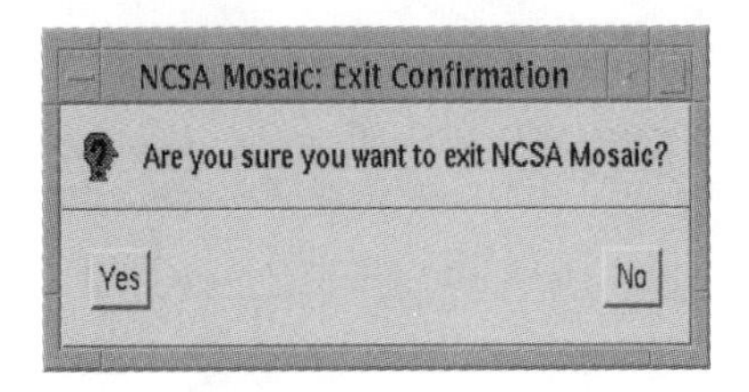

The Options Menu

You use the Options menu, shown here, to customize the appearance of the screen. Many of these menu items are toggles: selections with the two states on (as indicated by a check mark) and off. Each time you click on a toggle, you change its state. The Options menu offers nine selections, most of which are described in greater detail in later chapters.

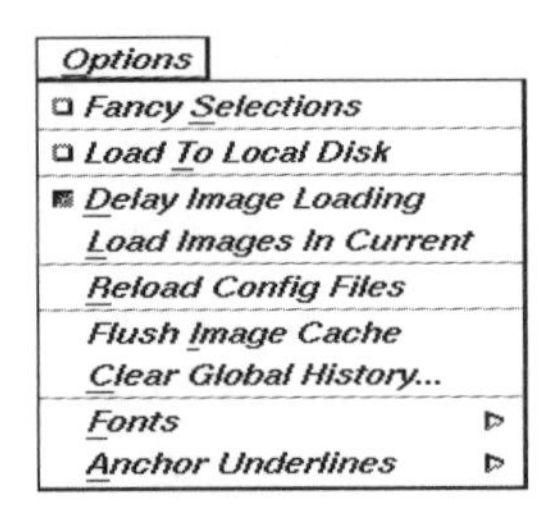

- **Fancy Selections**, when activated, preserves as much formatting (e.g., headers, tools) as possible with selections made from the X Window system cut and paste mechanism.

- **Load To Local Disk** enables you to specify that you want to save documents to a (local) disk.
- **Delay Image Loading** controls the simultaneity of text and image transmission. Turning this option on loads images after the text, which reduces transmission time for text data.
- **Load Images in Current** shows or suppresses images in the transmitted document. You can turn this option off to suppress images and reduce transmission time, especially if you're working on a busy network.
- **Reload Config Files** loads a copy of your previous configuration files. You can use this option to test different Mosaic configurations.
- **Flush Image Cache** removes presently saved (cached) images in order to reclaim system memory.
- **Clear Global History...** deletes the current history file, which can grow beyond manageable proportions in a busy system.
- **Fonts** allows you to select among numerous fonts.
- **Anchor Underlines** provides a choice of five underlining styles for hyperlinks.

The Navigate Menu

The Navigate menu, shown here, lets you maneuver efficiently across the Internet. The Navigate menu offers eight options, which are described next.

Navigate
Back
Forward
Home Document
Window History...
Hotlist...
Add Current To Hotlist
Internet Starting Points
Internet Resources Meta-Index

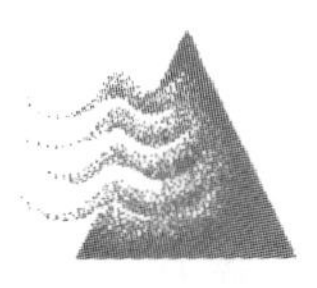

TIP
The time and energy you invest in mastering the Navigate menu will be paid back in spades.

- **Back** retrieves the most recently accessed document. Depending on how many documents you've retrieved in the current work session, you may be able to click on this item several times.

- **Forward** is the opposite of the Back menu item. Use this menu item when you press Back too many times and overshoot the desired document.
- **Home Document** returns you to your Home Page.
- **Window History...** displays an ordered list of the documents that you have accessed, as shown here:

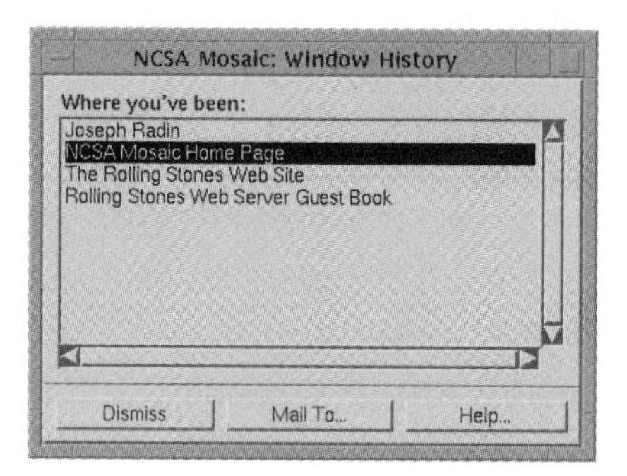

Select any item in the list and press the Load button to obtain it. Press the Dismiss button to close this window. Press the Mail button to send this information to NCSA Developers for help in debugging.

- **Hotlist...** lets you manipulate a Hotlist of documents for fast reference using the Hotlist View dialog box shown here:

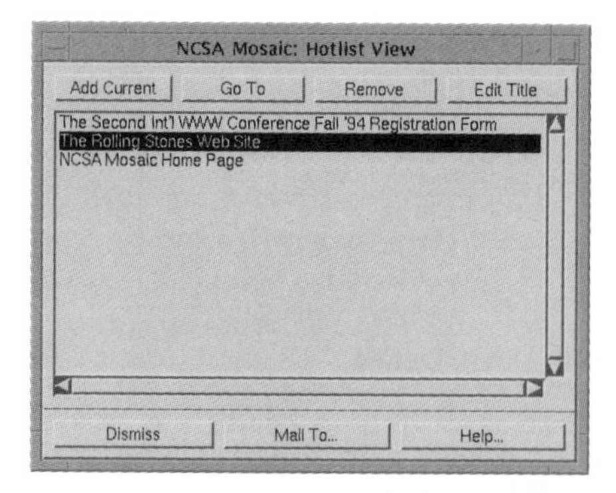

Click on the Add Current button to add a document to this list; click on the Go To button to immediately reference the selected document; click on the Remove button to delete a document from the Hotlist; click on the Edit button to change the Hotlist; or click on the Title button to give a title to the document.

- **Add Current to Hotlist** places the current document in a special list of documents to be retrieved often. Choosing this option is equivalent to clicking on the Add Current button in the Hotlist menu option.
- **Internet Starting Points** specifies where to start searching in the Internet, as shown in Figure 3-5.
- **Internet Resource Meta-Index**, shown in Figure 3-6, is a menu of search indexes for Internet resources. It includes indexes by subject and by server.

The Annotate Menu

You use the Annotate menu, shown here, to add your comments to Mosaic documents for future reference. The Annotate menu provides the selections described next.

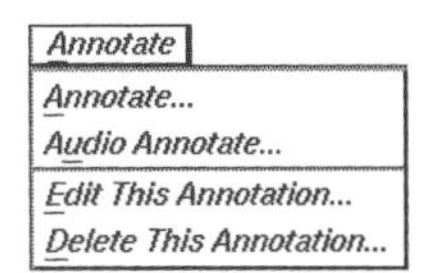

- **Annotate** generates the special window shown in Figure 3-7; here is where you compose your comments for the given file. You can enter *personal comments,* in which case only you can read them. You can also enter

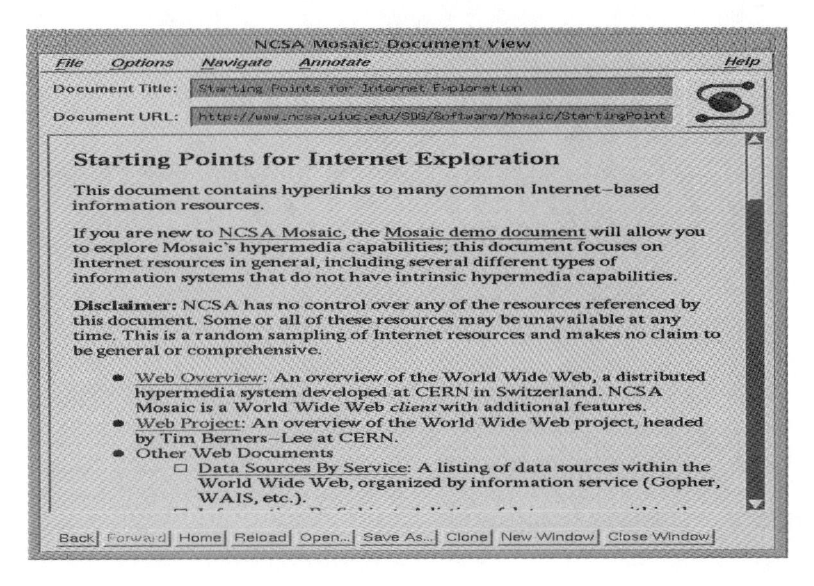

FIGURE 3-5. *Internet Starting Points*

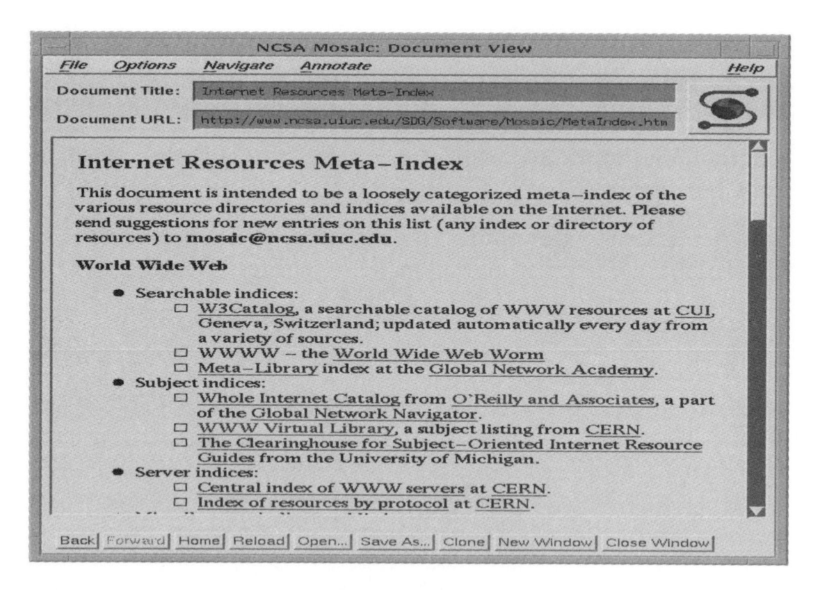

FIGURE 3-6. *Internet Resource Meta-Index*

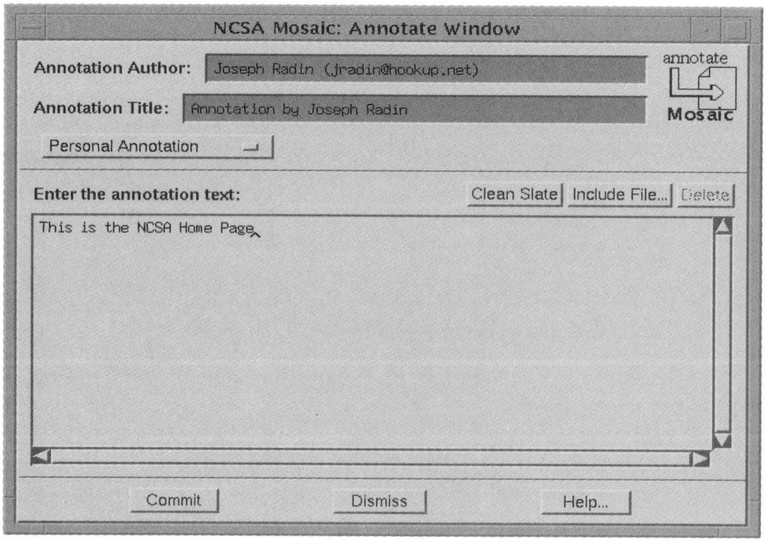

FIGURE 3-7. *Annotate Window*

group comments, which anyone in your workgroup can read. Finally, you can enter *public comments,* which anyone with access to the given document can read.

- **Audio Annotate...** opens the Audio Annotate window in which you can attach verbal comments to your document, provided that you have installed and configured the proper audio equipment. The annotation may be personal, workgroup, or public.
- **Edit This Annotation...** changes the contents of the designated annotation.
- **Delete This Annotation...** removes the designated annotation, permanently.

The Help Menu

You can use the Help menu, shown here, to obtain help when you need it. One reason for Mosaic's explosive growth is that its Help screens are so useful and so easy to find. From the Help menu you have a choice of ten selections, which are described in Chapter 8.

Help
About...
Manual...
What's New...
Demo...
On Version 2.2...
On Window...
On FAQ...
On HTML...
On URLs...
Mail Developers...

- **About...** presents brief information about Mosaic.
- **Manual...** provides an online manual.
- **What's New...** indicates some of the newest features of Mosaic.
- **Demo...** provides a short demonstration of Mosaic.
- **On Version 2.*x*...** describes the current version, comparing it to the previous version.
- **On Window...** explains the components of the NCSA Mosaic screen and menus.
- **On FAQ...** provides answers to frequently asked questions.
- **On HTML...** describes the Mosaic hypertext language.
- **On URLs...** describes Universal Resource Locators.
- **Mail Developers...** enables you to compose and send an information request to NCSA developers.

BUTTON	DESCRIPTION
Back	Displays the previous document in the history list.
Forward	Displays the next document in the history list.
Home	Generates the default Home Page.
Reload	Reloads the present document.
Open	Displays the Open URL dialog box and provides you with access to the Hotlist.
Save As	Saves the current document to disk.
Clone	Creates an additional copy of the current document window.

TABLE 3-1. *Mosaic for X Window (Unix) Buttons*

Mosaic Buttons for X Window (Unix)

Mosaic for X Window contains nine buttons at the bottom of the screen. Table 3-1 describes these buttons, from left to right.

Now that you've had an introduction to using Mosaic in X Window, you're ready for a guided tour.

A Guided Tour

This chapter provides a guided tour of Mosaic, showing just a few of its many possible uses. After establishing communications between your computer and the Internet, you will use Mosaic to explore a radio station's special programs, take a closer look at the cosmic event of the epoch, check out the latest information for golfers, examine the President's schedule, and see a movie about hurricanes. Along the way, you will discover how to use the main menu to control the screen image. The specific techniques you master here will serve you whether your interests are education, business, or just having fun.

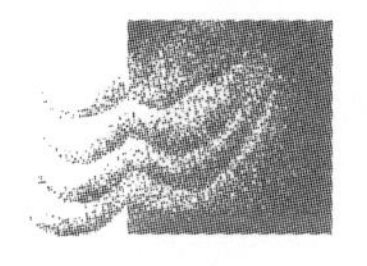

CAUTION

Mosaic is dynamic. New information sources are constantly going online and existing sources are continually updating and extending their information. So, don't be surprised if your screens differ from those in this book.

Most examples in this chapter apply to Mosaic for MS-Windows. You should already have loaded Mosaic as described in Chapter 2 and Appendix B. Mosaic for the X Window System (Unix), which was introduced in Chapter 3, provides a different window frame but essentially the same contents. We use Chameleon communications software and Hookup Communications as an Internet provider. Appendix B also presents Trumpet, another widely used communications software package.

Preliminary Steps

This chapter shows standard images and movie files for those who have installed the appropriate programs for seeing them (called *viewers*) following the instructions in Appendix B. If you haven't done this yet, do so now. No viewers, no images or movies. Before starting Mosaic, enter

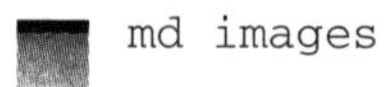

```
md images
```

to create a directory in which to store images on your hard disk. Then enter the command

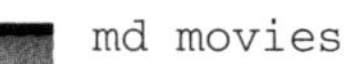

```
md movies
```

to create a directory in which to store movies on your hard disk.

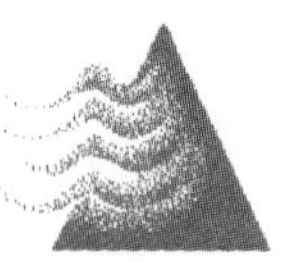

TIP
You won't be using these directories immediately, but nobody wants to interrupt a Mosaic session to do "housekeeping."

Until you have activated the appropriate communications software, Mosaic won't work. Figure 4-1 shows a sample screen generated while accessing the Internet via the communications software Chameleon for Windows, which includes a setup to connect to a provider, such as Hookup Communications. There is more information on providers in Appendix B.

Figure 4-2 shows the NCSA Mosaic for MS-Windows Home Page. As it stands, this document will not take you to your desired destination—namely, the list of commercial World Wide Web servers. To gain access to this list, you must specify its address, known as its Uniform Resource Locator (URL). Then you can save this useful address so you never have to enter it manually again.

Recall that the World Wide Web is a network of hypermedia servers. Any server may be reached from any other server. Mosaic is a client browser allowing

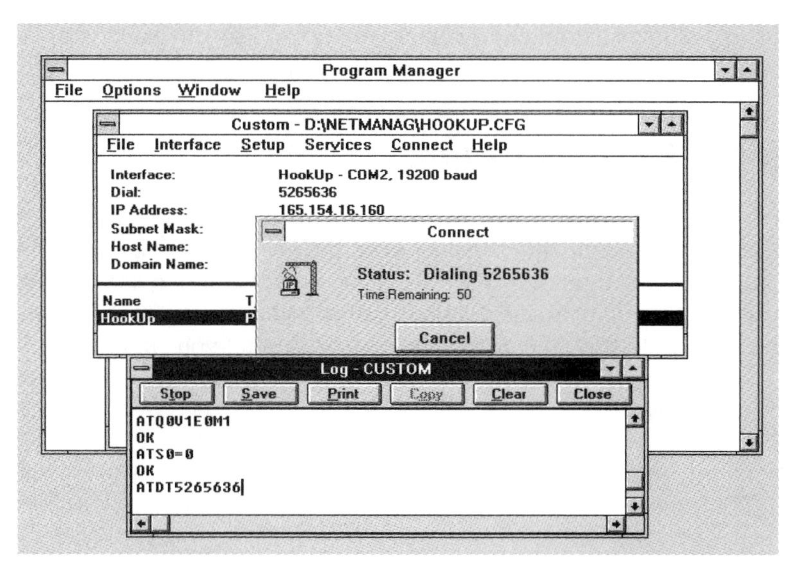

FIGURE 4-1. *Making the Mosaic connection*

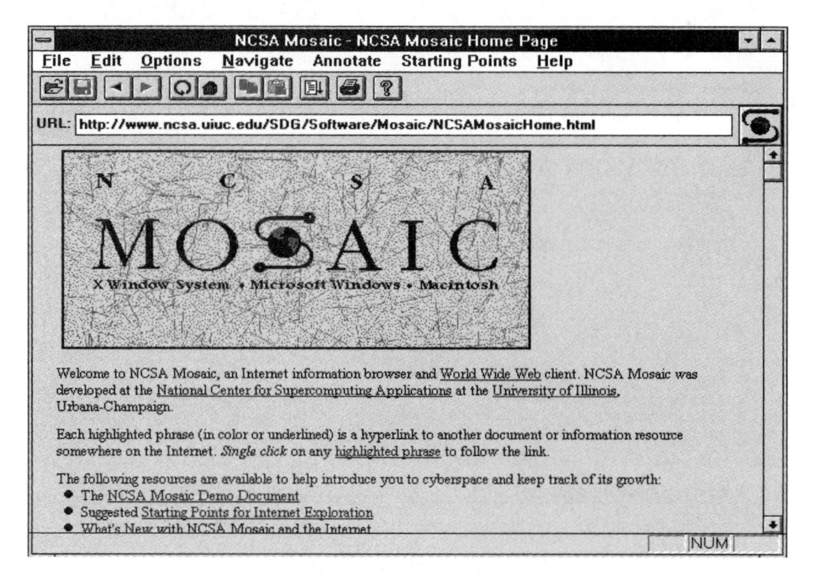

FIGURE 4-2. *The NCSA Home Page (MS-Windows version)*

you to view documents from any available server. You specify what you want and Mosaic accesses the appropriate server and furnishes you with the document.

Adding a Point of Interest

A point of interest is a shortcut for accessing the information you want. It is the Home Page for some entity (personal, commercial, or educational). Here you learn how to add a point of interest to your Mosaic system. Mastering this procedure for your version of Mosaic will save you lots of time and energy in the long run. If you don't know where to start, it will take longer to find what you want.

WIN

Select File from the main menu and then click on Open URL. Enter the following value, making sure not to make an error:

```
http://www.eit.com/web/www.servers/commercial.html
```

Press the OK key to confirm. This generates a screen similar to the one shown in Figure 4-3.

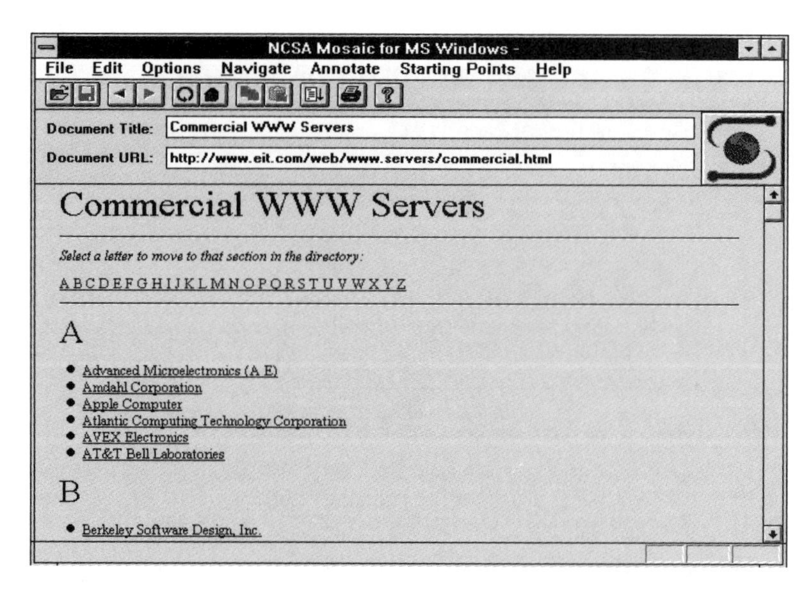

FIGURE 4-3. *Commercial World Wide Web Servers*

X

Select File from the main menu and then click on Open URL. Enter the following value, making sure not to make an error:

```
http://www.eit.com/web/www.servers/commercial.html
```

Select the Open option and then enter the document title Commercial WWW Servers. This generates a screen not very different from the one shown in Figure 4-3.

Before examining Figure 4-3 more closely, let's see how to add a document to the Hotlist and retrieve it without ever reentering the preceding URL. Once again, the necessary procedure is different for the two versions of Mosaic.

WIN

Select Navigate from the main menu and then click on Add Current to Hotlist. To retrieve the document, select File from the main menu and then click on the Open URL menu item. In the ensuing dialog box, click on the arrow to the right of the text box (upper-right) to generate a list of document titles, *or* click on the arrow to the right of the URL text box (upper-left) to generate a list of URLs. Select the proper document from either text box and click on the OK button to confirm.

X

Select Navigate from the main menu and then click on Add Current to Hotlist, or select Navigate, select Hotlist, and then click on the Add Current button in the ensuing dialog box. To retrieve the document, select Navigate and then select the Hotlist menu item. Double-click on the desired document from the displayed list or click on the document and then on the Go To button.

Figure 4-3 displays the first segment of the Commercial World Wide Web Servers list. All of the servers seem to be computer companies.

The Radio

Many people, ourselves included, listen to the radio while working on the computer. Frankly, we are getting a bit bored with our present station. To find a good alternative, scroll down through the present document (Commercial WWW Servers) to access the KKSF entry. After all, could a radio station in San Francisco be boring? Click on the underlined hyperlink in the present screen or the button to its left to launch the KKSF Home Page.

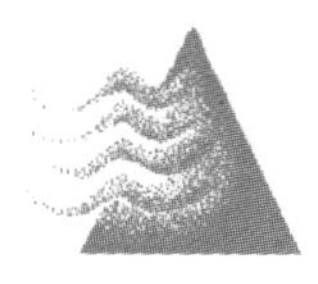

TIP
The spinning globe which appears in the upper-right corner of the screen informs you that document transmission is in progress. You can click on the globe at any time to stop transmission. Be patient; after several seconds you should see a screen resembling the one shown in Figure 4-4.

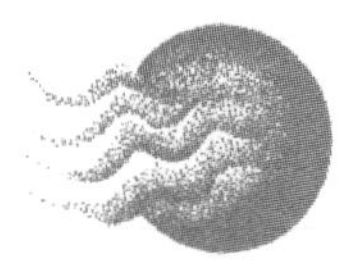

NOTE
It takes a lot longer to transmit text and images than plain text, because of the additional information. In this case, the images introduce us to KKSF and its sister classical station KDFC.

Do you like Beethoven and Bach? If you are in the listening area, head straight for 102.1 FM on your dial. Do you prefer Eric Clapton and Ray Charles? You could tune to 103.7 FM and try your luck.

The Collision of '94

Scroll down the page shown in Figure 4-4 until you get a screen similar to Figure 4-5, which provides information on the Comet Shoemaker-Levy 9 Jupiter Event! Let's take a look.

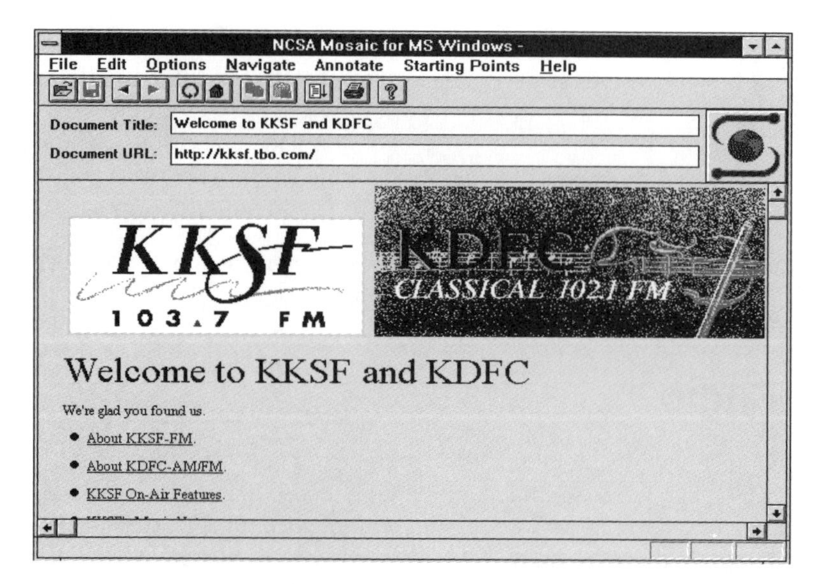

FIGURE 4-4. *A radio station Home Page*

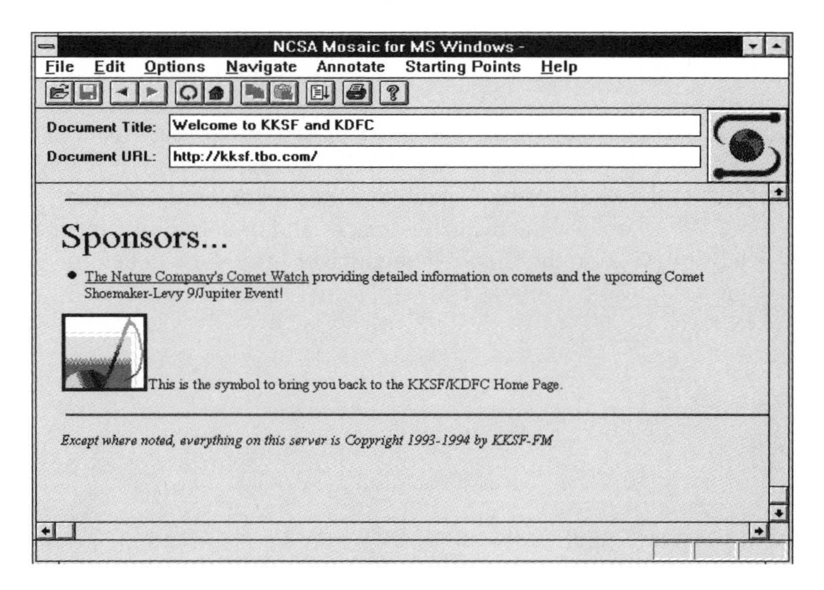

FIGURE 4-5. *Introducing a comet watch*

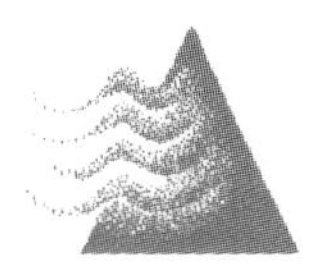

TIP
Click on the musical note icon to return to the top of the Home Page.

No drummer will ever produce as big a bang as the multiple collisions between the Comet Shoemaker-Levy 9 and the planet Jupiter in July, 1994. According to many, the last time the earth felt such a collision the result was a huge crater in southern Mexico and the extinction of the dinosaurs. In any case, the Comet-Jupiter collision was something to see. And, through the miracle of Mosaic, you can access it via the radio by clicking on the hyperlink The Nature Company's Comet Watch.

The document URL is

```
http://kksf.tbo.com/nc/nc.html
```

clearly associated with the previous Home Page whose URL is

```
http://kksf.tbo.com
```

Unfortunately we are late for the radio station's contest testing your comet trivia knowledge. Scroll down to the screen shown in Figure 4-6 to see what you could have won.

The icon at the bottom of Figure 4-6 has a colored border, signifying that it's a hyperlink. This icon also appeared in Figure 4-5. Clicking on this musical note brings you back to the KKSF Home Page. Don't touch that icon! Let's gaze at some crash pictures. We examined the available images and preferred the one that you can generate by clicking on the Comet P/Shoemaker-Levy 9 Impact Home Page at the University of Arizona hyperlink. The results are shown in Figure 4-7.

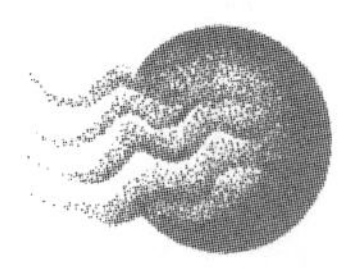

NOTE

This figure's Document URL is

```
http://seds.lpl.arizona.edu/sl9/sl9.html
```

which is totally different from the URL for Figure 4-6 that showed the KKSF Home Page.

Changing the Home Page changes the URL, and consequently changes the server.

As you already know, Mosaic is associated with the World Wide Web. Like a spider web, the Web offers many paths from one information point to another. Instead of taking the path we did, we could have reached the screen shown in Figure 4-7 via the University of Arizona Home Page. Doing so requires that we know the appropriate URL, listed in Appendix A, and that we use the procedure for adding a starting point presented in the beginning of this chapter. Above all, you have to know that the University of Arizona is an excellent place to look for information about the comet watch. The beauty of hyperlinks is that they enable you to find information easily that otherwise you may not have known to exist. Let's see what else we can get out of this Solar System Smash-up.

Scroll down the Comet P/Shoemaker-Levy 9 Impact Home Page, mercifully abbreviated the SL9 Home Page. Click on the Images and Animation hyperlink, and then click on Frames from a computer simulation.

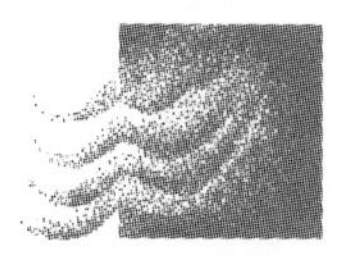

CAUTION

The wording for different hyperlinks may be similar. In fact, four of the eight entries in the bulleted list include the phrase "computer simulation." Read each entry carefully before clicking on the one you want.

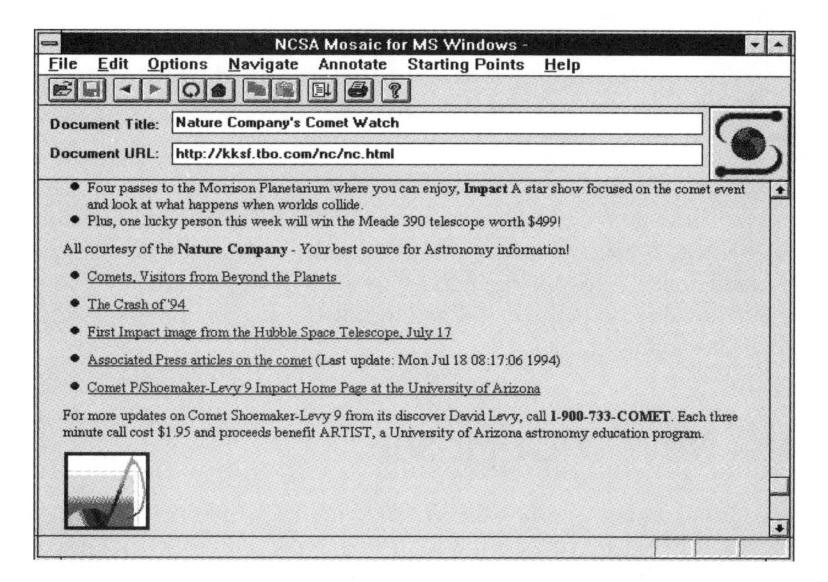

FIGURE 4-6. *Listing comet watch options*

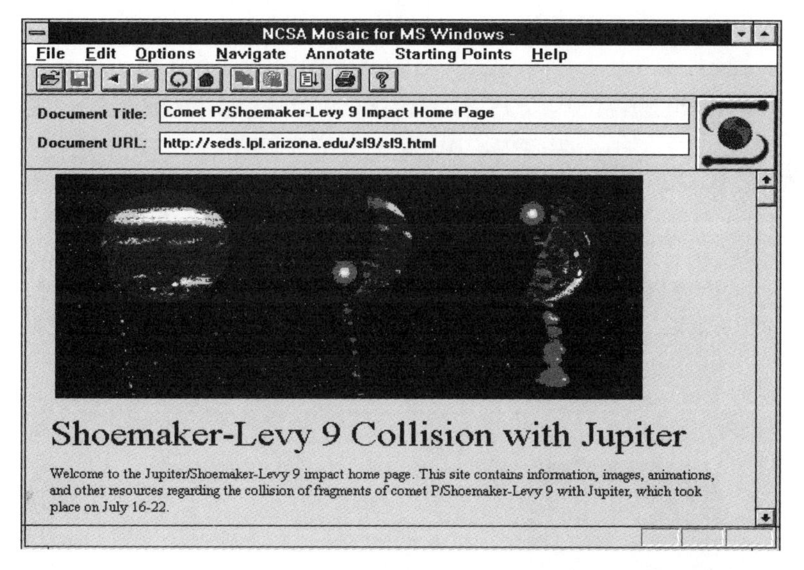

FIGURE 4-7. *The collision of the epoch*

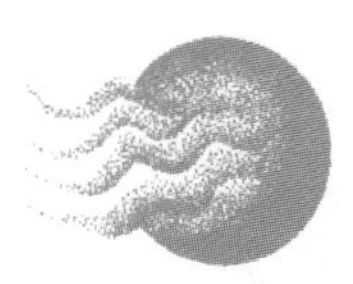

NOTE
As mentioned, Home Pages are often updated. Don't be surprised if the information in the SL9 Home Page changes.

Figure 4-8 shows two close-up shots of the collision between Jupiter and the comet. Each image is accompanied by two buttons. The magnifying glass zooms in on the selected image. The tape reel icon shows a movie, provided that you followed the directions in Appendix B and installed a Mosaic movie viewer. You'll see a movie later in this chapter.

What's New with Mosaic

New servers are constantly joining the World Wide Web, and new information is added to existing servers. Even if last week's search proved unsuccessful, you should look again. The information you need may have come online yesterday. And the best thing is that you don't have to repeat the entire search process. You can go back to selected documents as required by clicking on the Back icon in the toolbar.

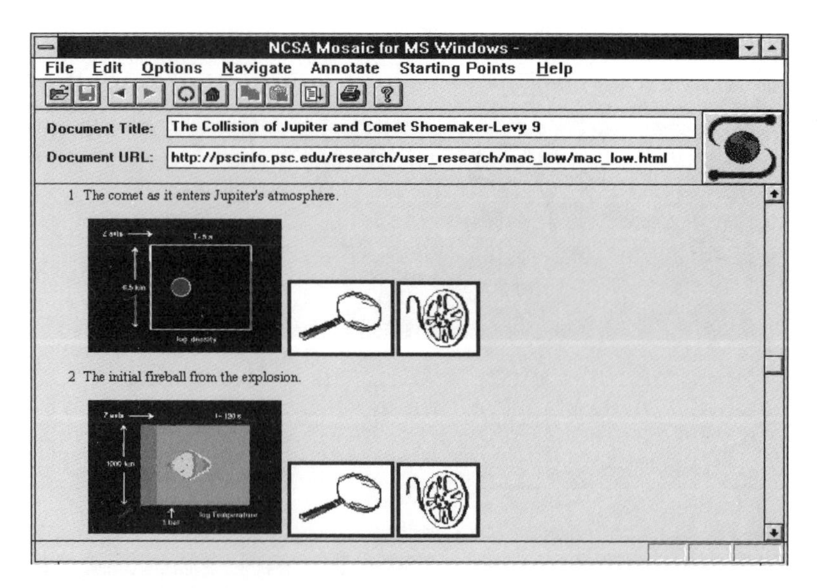

FIGURE 4-8. *Simulating the collision*

WIN

Return to the NCSA Mosaic Home Page by clicking on the House icon in the toolbar. Then select Starting Points from the main menu.

X

Return to the NCSA Mosaic Home Page by clicking on the Home button at the bottom of the screen. Then select Navigate and click on Internet Starting Points.

Once you have accessed the starting points, select the NCSA Mosaic's What's New Page. This generates the What's New With NCSA Mosaic and the WWW screen shown in Figure 4-9.

The What's New document is several screens long. After a brief introduction telling you how to use this document, the screen includes information starting with the most recent date first. We accessed this document on the evening of July 27, 1994. It included updates posted the previous day. Scroll down in the What's New document. We browsed the night before Joseph's annual company Golf Day, so he clicked on the GolfData On-Line's Home Page hyperlink in the hope of finding something—anything—that would help his game.

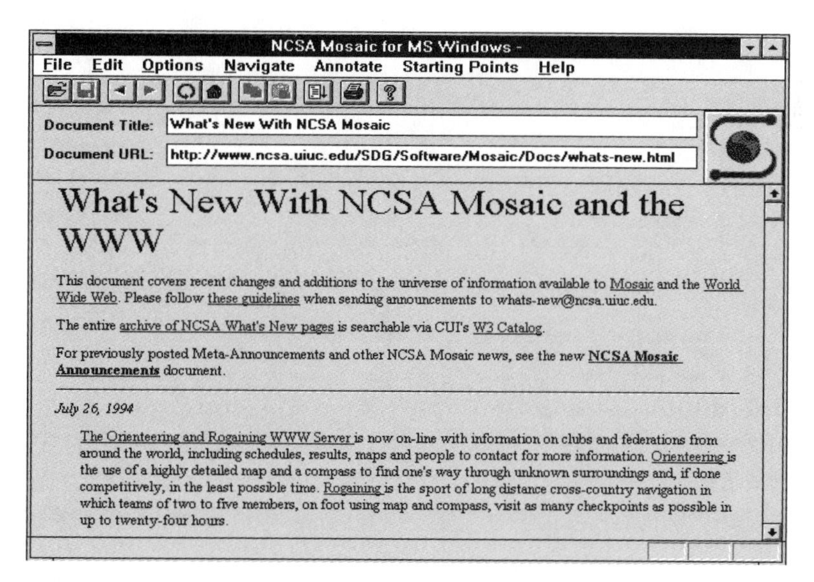

FIGURE 4-9. *Checking the latest updates*

Golfing

Figure 4-10 shows the GolfData On-Line Home Page. Its URL is

```
http://www.gdol.com/
```

We clicked on the golf ball to the left of Known Golf Links. Would Joseph's course be listed? If so, would the document suggest how to stay out of the sand traps? The answer is No. Figure 4-11 shows the Known Golf Links document. Alas, Joseph was not going to play in Princeton, Dartmouth, Alberta, or Glasgow. But the 19th Hole is always interesting, especially after a long game or surfing session. So let's click on it.

It took several minutes to generate Figure 4-12. However, the wait was worth it. The image was spectacular. Many information providers supply top-quality images. We scrolled down the document to generate Figure 4-13. Note the special icon to indicate an area under construction, for which your suggestions are solicited.

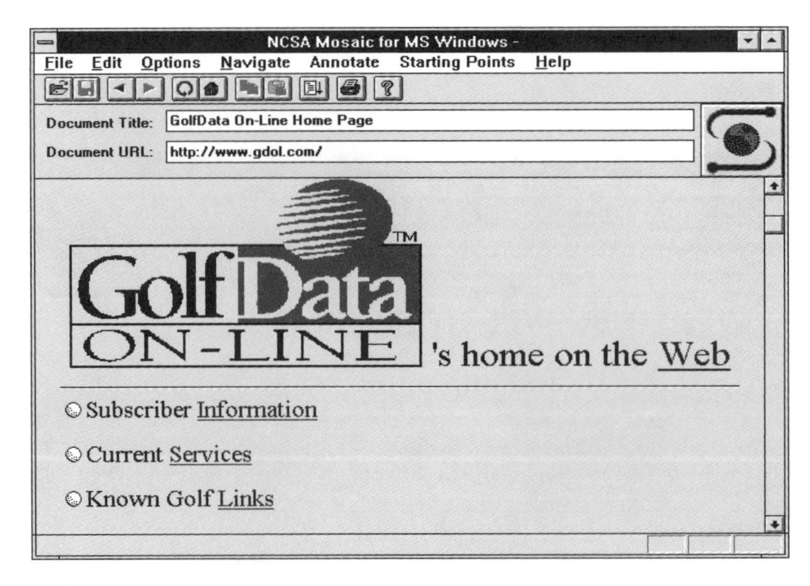

FIGURE 4-10. *The GolfData On-Line Home Page*

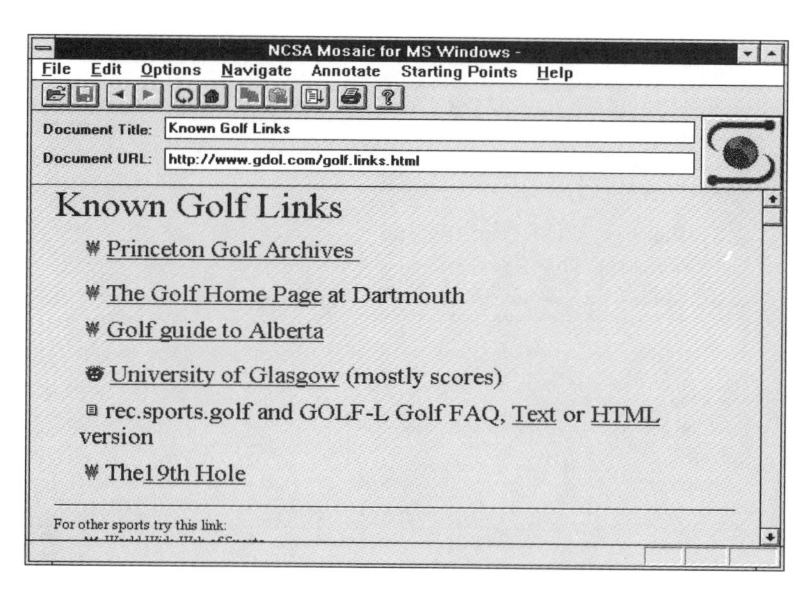

FIGURE 4-11. *Are these the only known golf links?*

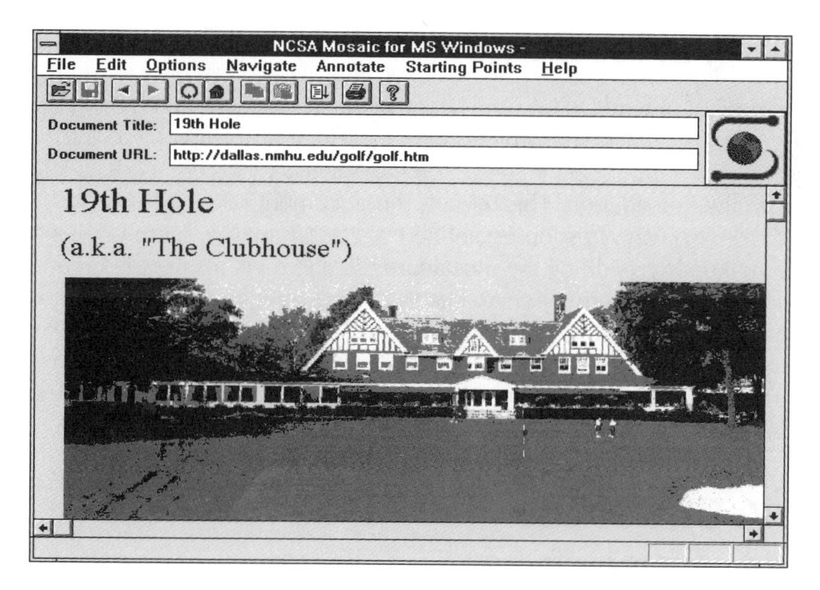

FIGURE 4-12. *What a clubhouse!*

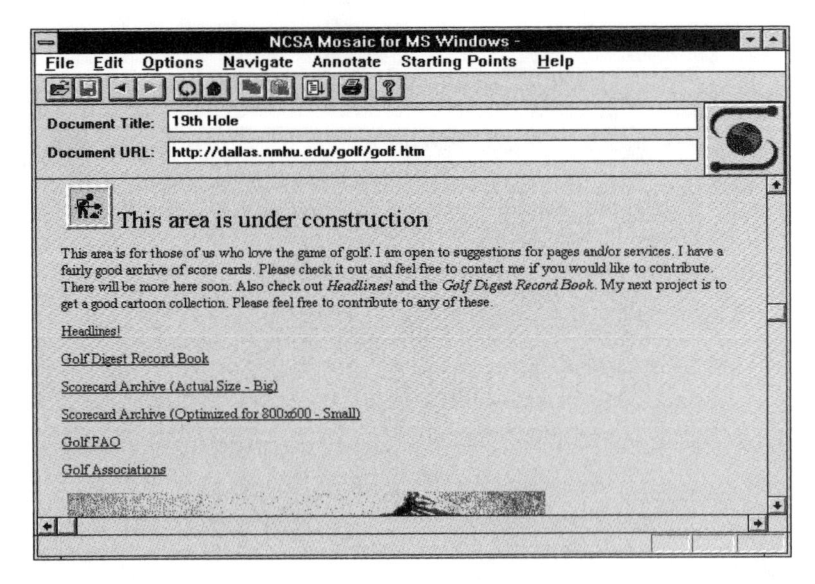

FIGURE 4-13. *Still under construction*

Disabling Inline Images

Image transmission tends to be slow because images take up much more disk space than text. You can reduce transmission time substantially (often by a factor of ten or more) simply by suppressing images. You can always retrieve the image later by clicking on the Reload icon. This reloads the document.

There are two ways to suppress inline images. Chapter 6 demonstrates how to do so permanently by editing the **mosaic.ini** file. Here we access the Options menu item in the main menu and click on Display Inline Images (Delay Image Loading for X Window systems) to disable these images. This generates a screen like the one shown in Figure 4-14, in which images are replaced by the Mosaic logo with the word *Image* under it. Using this procedure reduced the document transmission time from a few minutes to a few seconds.

In and Out of the White House

1600 Pennsylvania Avenue—the very address reeks of power. This next series of exercises seeks information on this location and its inhabitant, the President of the United States.

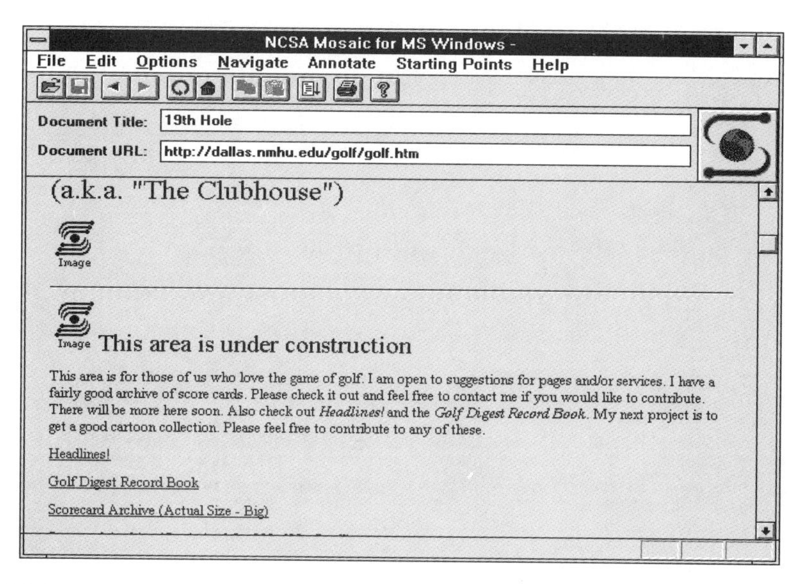

FIGURE 4-14. *The image is gone*

Return to the NCSA Home Page by clicking on the Home icon in the toolbar. Access Carnegie Mellon using the following URL.

```
http://www.cs.cmu.edu:8001/Web/FrontDoor.html
```

or carry out the following procedure.

WIN

Click on the Starting Points selection in the main menu, then click on the Home Pages selection and select the Carnegie Mellon option to generate the Carnegie Mellon School of Computer Science Home Page, which is shown in Figure 4-15.

X

Click on the Navigate selection in the main menu and then click on the Internet Resources Meta-Index. Select the *Central Index* of *WWW Servers* hyperlink and select Carnegie Mellon University.

Clicking on the hyperlink other CMU servers generates a screen associated with the central CMU server. Then click on the English Server hyperlink toward the bottom of the screen to generate the screen shown in Figure 4-16. The English

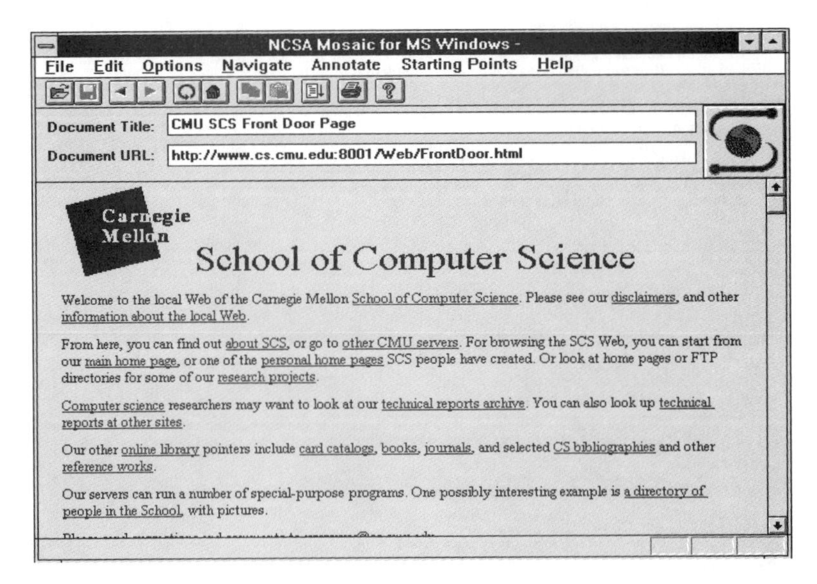

FIGURE 4-15. *The CMU School of Computer Science Home Page*

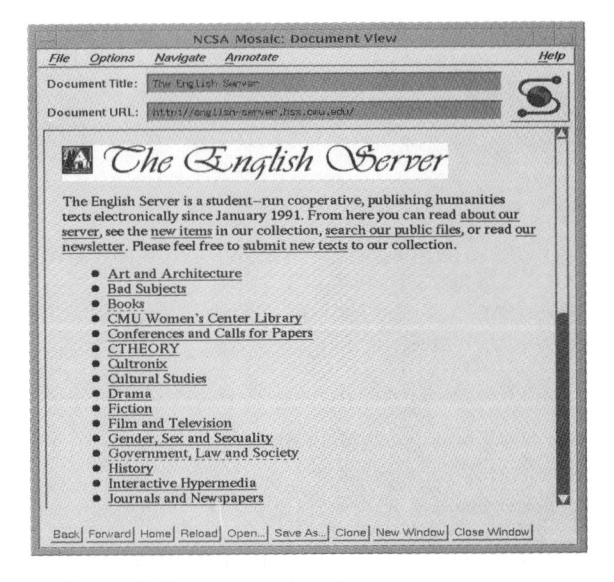

FIGURE 4-16. *The English Server Home Page*

Server is a student-run cooperative that has been publishing humanities texts electronically since January 1991.

Click on the Government, Law, and Society hyperlink. (You'll have to check out the Gender, Sex and Sexuality hyperlink yourself.) There's nothing about the Presidency here. Don't despair. Scroll down until you see the Gopher hyperlink and click on it to generate a screen like the one shown in Figure 4-17. The first line of text indicates that the information comes from a Gopher menu. (Don't think of Gopher as a competitor to the World Wide Web.) As you may recall from Chapter 1, Gopher is a menu-driven Internet server that was developed at the University of Minnesota. As this figure shows, Mosaic can access information from Gopher servers.

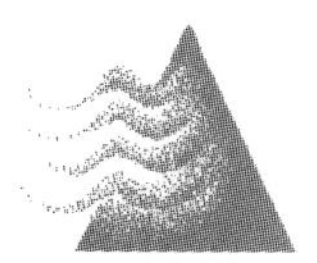

TIP

You can also access this document by choosing Open URL from the File menu and entering the URL **gopher://gopher_address** in the Document URL box. This process will be explained further in Chapters 5 and 7.

Click on the hypertext link for the Public Schedule for June 20, 1994 or the icon to its left. This generates the President's public schedule for that date. Once again his schedule did not include us.

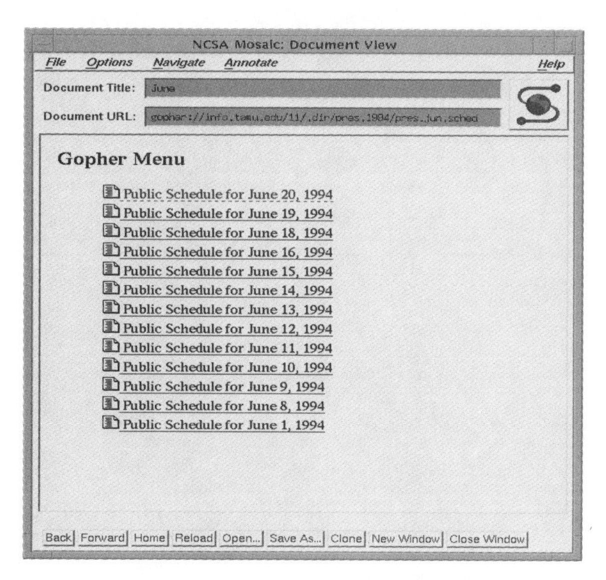

FIGURE 4-17. *The President's schedule via Gopher*

If you want to see a lovely image of the White House and access associated information, you can return to the English Server and start searching. Alternatively, you can access the document manually with the following document URL:

```
http://english-server.hss.cmu/WhiteHouse.html
```

The resulting screen is shown in Figure 4-18.

This concludes our guided tour of the White House, but, now that you have accessed the White House Files you can continue on your own. We're going to Honolulu. Don't get too excited; we are going for a look at hurricanes.

Weathering a Hurricane–Looking at Images and Movies

Next you'll see a series of hurricane images taken in Honolulu, Hawaii. Don't go any further until you have installed the image and movie viewers according to instructions provided in Appendix B. Once you've done this, you will be able to see high-quality images and watch a hurricane movie.

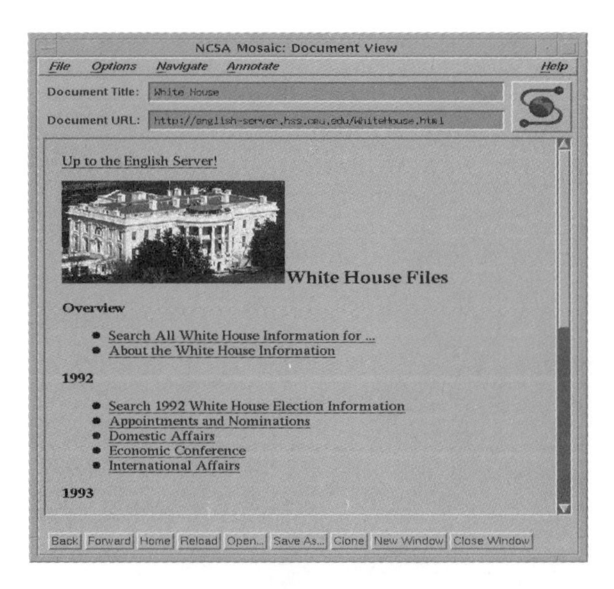

FIGURE 4-18. *The White House Files*

WIN

Choose Options from the main menu and enable the Load to Disk option.

X

Choose Options from the main menu and enable the Load to Local Disk option.

Turning on the Load to Disk option causes incoming files to be stored to disk, instead of being transferred into memory. You can then access these files at your leisure, instead of having to deal with them immediately. There are other advantages to storing files to disk. You can view them repeatedly and process them with other programs. For example, you could piece together several movie files, provided that you have the legal permission to use them.

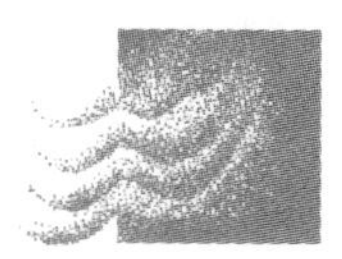

CAUTION

Recall that turning the Load to Disk option on surpresses the display of the file. Turn this option off whenever you don't need it.

Return to the NCSA Home Page and then create a starting point for Honolulu Community College using the following URL. (If you don't remember how to create a starting point, refer to the section "Adding a Starting Point" toward the beginning of this chapter.)

```
http://www.hcc.hawaii.edu/
```

This generates the Honolulu Community College Home Page shown in Figure 4-19. Before seeing a movie, you will experiment with a few menu selections that modify the screen presentation. These menu selections are available with the MS-Windows version of Mosaic.

Modifying the Screen Presentation

The following menu selections help you gain a little extra elbow room on the screen. First, you will remove the toolbar from the screen.

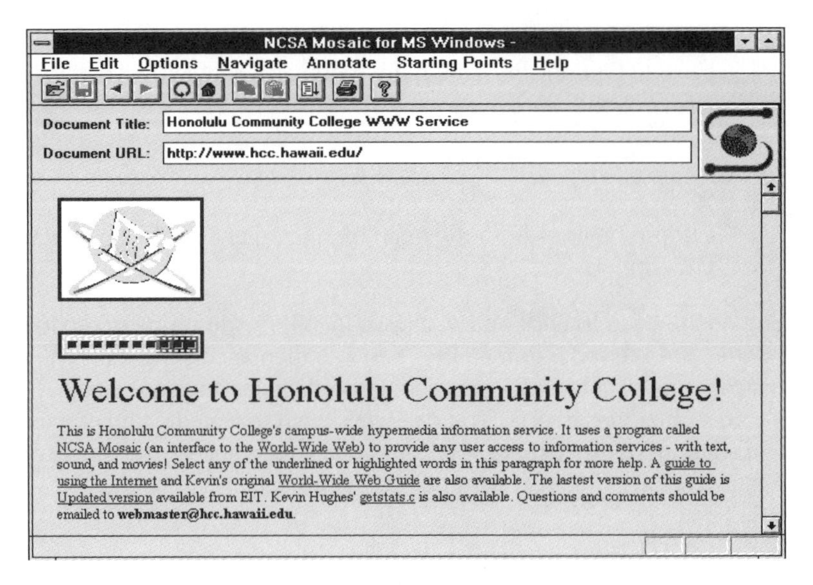

FIGURE 4-19. *The Honolulu Community College Home Page*

WIN

Select Options and then click on Show Toolbar, removing the checkmark beside it. This suppresses the toolbar, which normally appears near the top of the screen. The modified display is shown in Figure 4-20.

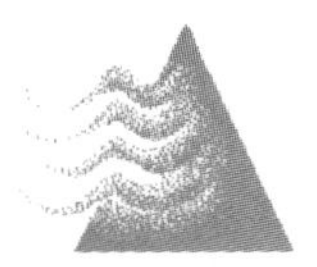

TIP

It's a good idea to keep the toolbar on. It provides a quick way of doing useful things such as returning to the previous document (left arrow icon), reloading a document (circle icon), returning to the Home Page (house icon), and printing the current document (printer icon).

You can also remove from the display the URL and title associated with the current document.

WIN

Select Options and then click on Show Current URL/Title, removing the checkmark beside it. This suppresses the document title and URL near the top of the screen. The modified display is shown in Figure 4-21.

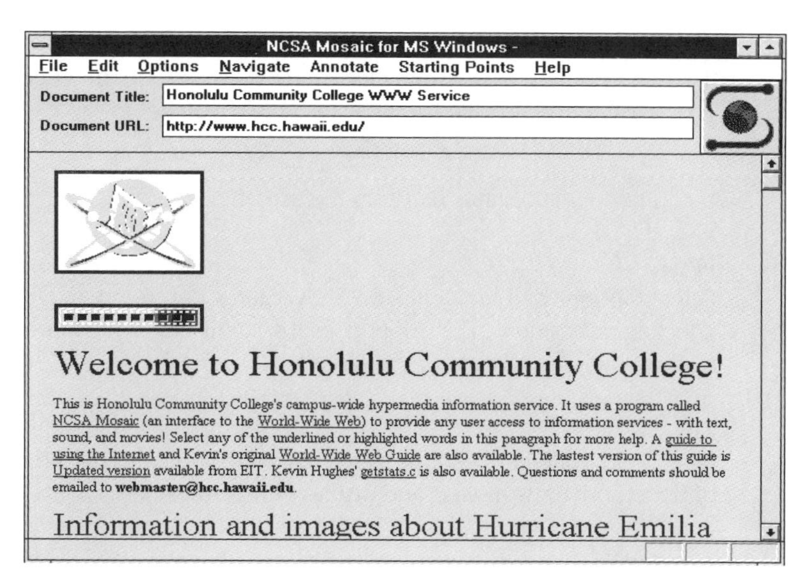

FIGURE 4-20. *A screen with no toolbar*

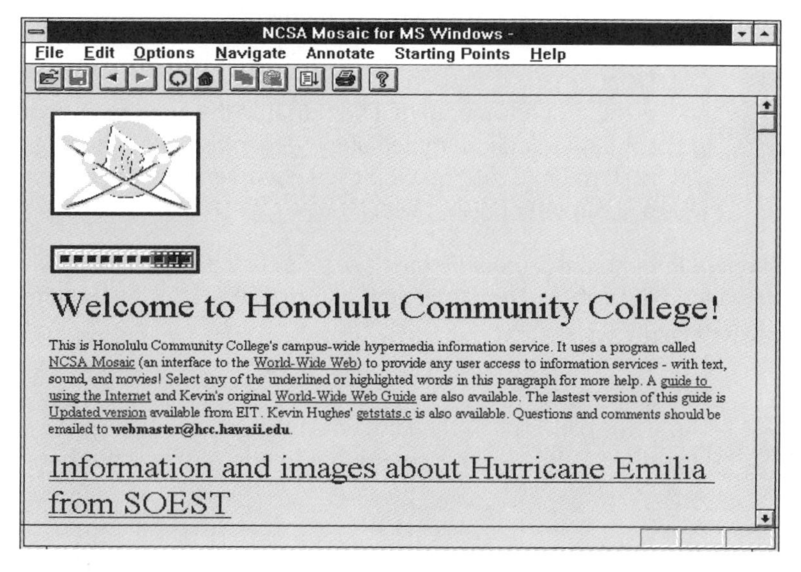

FIGURE 4-21. *A screen with no document title or URL*

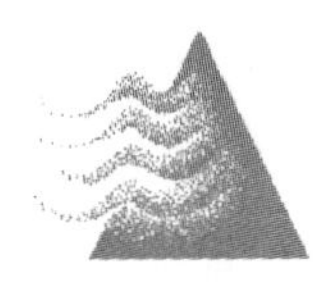

TIP
It's a good policy to keep the Document Title/URL option checked. This information is quite useful, especially if you get lost.

Finally, you can remove the status bar from the screen.

WIN
Select Options and then click on Show Status Bar, removing the checkmark beside it. This suppresses the status bar, which normally appears near the bottom of the screen.

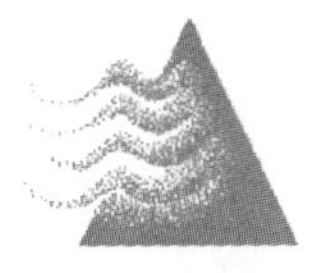

TIP
It's a good policy to keep the status bar on. It indicates the progress of document transmission and displays error messages.

Processing Movies and Special Images

Click on the film strip icon above the word *Welcome* in the Honolulu Community College Home Page to display a pop-up menu requesting the name of the local disk file that will store the incoming movie. Accept the default file name by clicking on OK. If no errors occur, you will be notified when the file transfer to disk is complete. We'll take a look at this movie in a moment.

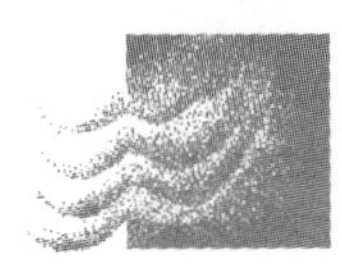

CAUTION
Access the Options menu item and disable the Load to Disk or Load to Local Disk selection immediately. Otherwise, all incoming files will be stored to disk. You won't see documents on the screen and you may run out of disk space.

Scroll down through the present Home Page and click on the hyperlink Information and images about Hurricane Emilia from SOEST. You will see the index shown in Figure 4-22.

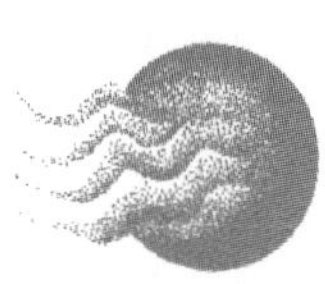

NOTE
The status bar displays the URL when the cursor is over the hyperlink. This is another reason not to turn off the status bar.

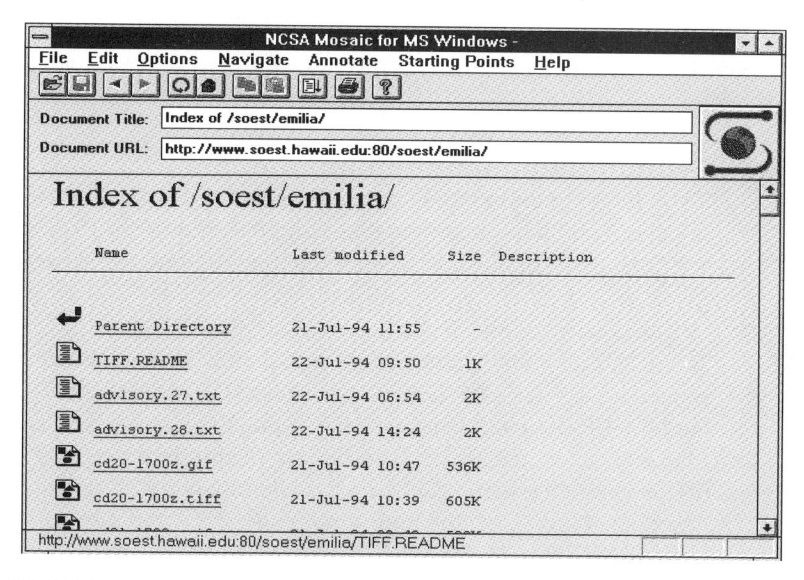

FIGURE 4-22. *An index of hurricane files*

Viewing Movies and Special Images

Unfortunately, the Description column shown in Figure 4-22 is blank. We may have to access the documents to see what they contain. We could click on TIFF.README to get more information. Instead, let's plunge right into a gif file. *gif files* contain images in the CompuServe format. These files tend to be big and consequently take a long time to transmit. This shouldn't be suprising given the wealth of detail that they contain. To access one use the appropriate procedure described next.

WIN

Choose Options from the main menu and enable the Load to Disk option.

X

Choose Options from the main menu and enable the Load to Local Disk option.

Click on the hyperlink for the cd20-1700z.gif file. The file size is 536K bytes; expect to wait about 4 to 6 minutes for transmission over a medium-speed line.

Once the image is transferred to hard disk (in our case the **images** directory) disable the Load to Disk or Load to Local Disk selection in the Options menu before viewing the file.

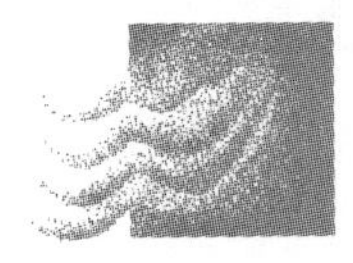

CAUTION

If you forget to disable the Load to Disk or Load to Local Disk selection, all future incoming files will be stored to disk. You won't see documents on the screen and you may run out of disk space.

WIN

To see this gif image you must activate a viewer. The following procedure activates the L31 viewer, defined as the default viewer in the NCSA Mosaic configuration file. Open the Windows File Manager, select the appropriate disk drive and directory, and click on the image file name. Click on the File menu option and then on Associate. Next select the appropriate disk drive and directory and click on the **lview31.exe** file. Click on OK twice to confirm. Whenever you double-click on the image file name, the viewer is activated and you'll see the special image as shown in Figure 4-23. Appendix B explains how to install the viewer of your choice.

FIGURE 4-23. *A special hurricane image*

X
To see this gif image you can use the xv viewer, whose installation is discussed in Appendix B.

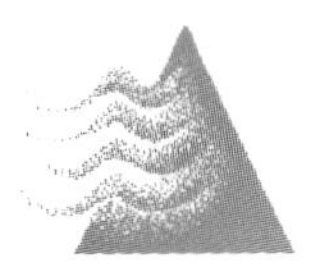

TIP
To save time note the location of the image file and the viewer file before associating them.

Follow a similar procedure to see the movie via the Mpegplay viewer (file name **mpegplay.exe**). Recall that movie files have the extension .mpg and are stored in a directory such as **movies**. Our hurricane movie is shown in Figure 4-24.

This concludes the guided tour of Mosaic. In Chapter 5 we'll explore more searching techniques and strategies for Mosaic.

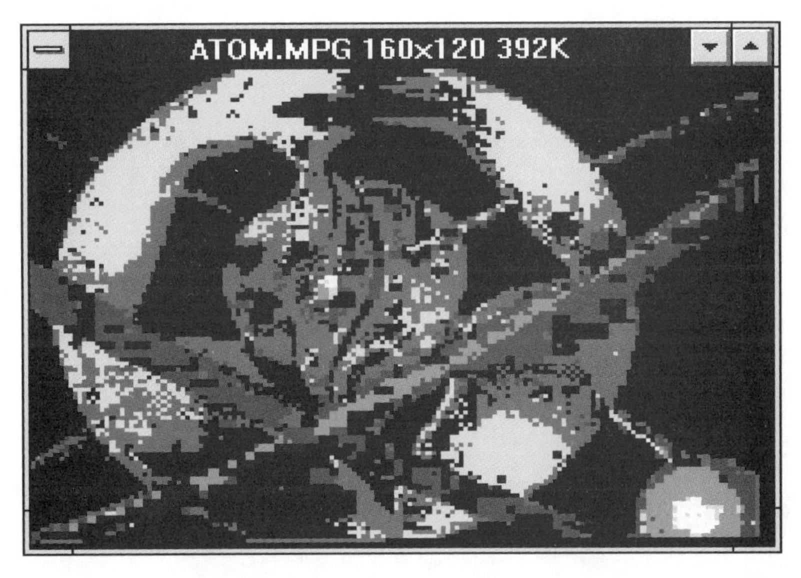

FIGURE 4-24. *A hurricane movie still*

Searching for Information with Mosaic

Now that you have finished touring Mosaic, it's time to examine the techniques and strategies for conducting information searches with Mosaic. This chapter explains how to get the information you need efficiently. First you learn a bit more about Uniform Resource Locators and Hypertext Markup Language. At the beginning of every information search is a Home Page, created and maintained by an individual, a company, or a nonprofit organization such as an educational institution. This chapter examines in detail the similarities of and differences between these categories of Home Pages. Part of Mosaic's strength is its ability to work with other systems to access the Internet. You learn how to retrieve information stored on Gopher servers via Mosaic. You can even get news from the Internet. Once you find the information you want, you can download it through Mosaic. The chapter concludes with an examination of WAIS, another Internet searching tool that you can access with Mosaic.

Two Important Mosaic Acronyms

Every technical subject has its acronyms, such as WWW (World Wide Web). Two commonly used Mosaic acronyms are URL (Uniform Resource Locator) and HTML (Hypertext Markup Language). As you will see in this chapter, these two subjects are closely associated with the Home Page.

The URL is the standard way of identifying documents on the World Wide Web, including Home Pages. HTML is the programming language used to create World Wide Web documents, including Home Pages. These two subjects are discussed next. The Hypertext Markup Language is covered in greater detail in Chapter 6.

Uniform Resource Locator (URL)

The Uniform Resource Locator (URL) is the standard addressing mechanism used to locate and retrieve documents anywhere on the World Wide Web. When you click on a hyperlink in the current document, Mosaic retrieves the document whose file name is given by the URL. Most users don't know what's going on behind the scenes; they simply click and wait for the linked document to be displayed.

A URL consists of three parts:

- A code identifying the transfer protocol to be used.
- An address identifying the server on which the file resides.
- A full path name locating the file on that server.

Consider the URL for the NCSA Home Page:

```
http://www.ncsa.uiuc.edu/SDG/Software/WinMosaic/HomePage.html
```

The first part, **http://**, identifies the transfer protocol, namely, the Hypertext Transfer Protocol, specifically designed for the interactive, hypermedia WWW environment. The second part, **www.ncsa.uicu.edu**, as read from left to right, identifies the World Wide Web server for the NCSA at the University of Illinois at Urbana-Champaign, an educational server. The third part, **SDG/Software/WinMosaic/HomePage.html**, identifies the directory and file name.

URLs are not restricted to the http protocol. Let's look at file, ftp, gopher, and news URL examples. First, consider a file URL such as the following:

```
file://ftp.myserv.com/mydir/files/mydoc.txt
```

The file **mydoc.txt** is associated with the ftp server **ftp.myserv.com** in the directory **/mydir/files**.

Next consider the general form for an ftp URL:

```
ftp://ftp_server_name/directory_and_file_name
```

The acronym ftp, introduced in Chapter 1, stands for file transfer protocol, a service for transferring files from one computer to another. Appendix B shows how to use the ftp utility to install Mosaic and related software.

Next consider the general form for a Gopher URL:

```
gopher://gopher_server_name/directory_and_file_name
```

Gopher, introduced in Chapter 1, is a powerful, menu-based system for accessing information from the Internet. Later in this chapter, you learn more about accessing Gopher files via Mosaic.

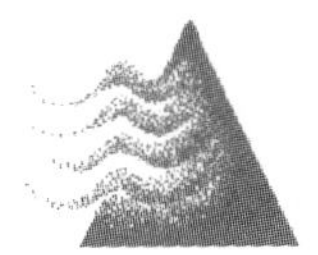

TIP

If you know that the designated Gopher server mygoph is on a given port instead of the default network port, use the following URL:

```
gopher://mygoph/directory_and_file_name:gopher_port/
```

Finally, consider a news URL:

```
news:newsgroup.subject
```

USENET News, also called Net News, is an interactive Internet service bringing the latest news to millions of subscribers. Given the volume of available news, subscribers must indicate their area of interest by subscribing to particular newsgroups, of which there are now about 4,000.

Newsgroup names are hierarchical, reading from left to right. As you move from left to right, the categories become more specific. The newsgroup **news** contains information about USENET. The newsgroup **news.software** contains information about software for USENET. And the fictitious newsgroup **news.software.readers.lefties** contains information about news readers for the left-handed.

The following URL accesses the newsgroup name for real and model trains:

```
news:rec.railroad
```

Now that you know how URLs are constructed, you can apply them without fear.

Hypertext Markup Language (HTML)

The *Hypertext Markup Language (HTML)* is a specialized programming language used to create hypermedia documents stored on the World Wide Web and browsed via Mosaic. HTML documents consist of text, formatting codes, and hyperlinks to other documents. Some elements describe the entire document or link it to other documents. Other elements format portions of the text. Still others insert graphics within the document.

The next section discusses the different categories of Home Pages, illustrating the use of elementary HTML features in the context of a simple Home Page. Chapter 6 presents the details of HTML syntax and introduces a very useful software package for automating the creation of HTML documents.

Different Types of Home Pages

Home Pages fall into two basic categories: relatively simple personal Home Pages and more complicated corporate and educational Home Pages. Recall from Chapter 1 that Mosaic is based on the client-server model, which allows resource sharing on a network. This sharing is handled by independent programs running on different computers. The server program provides the resource, such as a document. The client program accesses the resource. You'll see soon how the different types of Home Pages are associated with the client-server model.

Most users don't think about the type of Home Page they are using. During a given search they may access a wide variety of Home Pages. However, there are definite advantages to starting a Mosaic session from a personal Home Page rather than a corporate or educational Home Page such as the widely used NCSA Home Page. These advantages are as follows:

- Selecting the document takes fewer steps. You can customize your Home Page to meet your specific needs. Suppose you often require the Commercial World Wide Web Servers list. From the NCSA Home Page, you must first click on Starting Points and then select Commercial World Wide Web Servers from the displayed list. But from the customized Home Page shown in Figure 5-1, you need only click on the hyperlink WWW Servers.
- System access to the document is faster. It takes much less time to access a Home Page stored on a local disk than to access a Home Page stored on a remote disk.
- The load on the server containing the remote, shared Home Page is reduced. For example, the load on the NCSA Web server is so great that users have been formally requested not to load an NCSA Mosaic Home Page automatically at system startup. In contrast, the NCSA server won't even know when you access your own Home Page.
- Availability is increased. When the remote server for your Home Page is not operational, you cannot use it until they fix it. When your computer for your local Home Page is not operational, if you can't fix it yourself, you can at least call a technician.

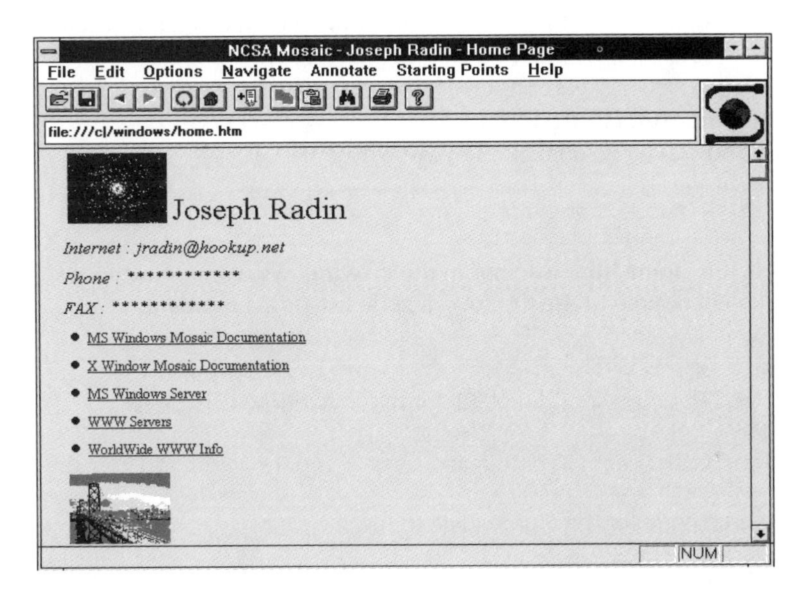

FIGURE 5-1. *A personal Home Page*

Personal Home Pages

A personal Home Page is created and maintained by an individual user running Mosaic as a client. In general, a personal Home Page is relatively short, perhaps one or two screens long. This Home Page is accessed by one person, the user who created it. Because only one person accesses a given personal Home Page, errors tend not to be critical; for example, a URL may be incorrect. A personal Home Page is fairly static, evolving slowly over time.

WIN
HTML documents (files) have the extension htm.

X
HTML documents (files) have the extension html.

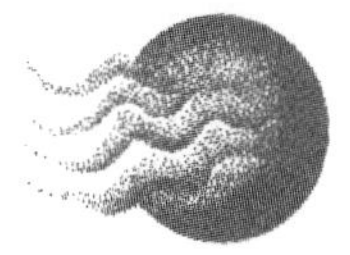

NOTE
The personal Home Page shown in Figure 5-1 was created to run under the MS-Windows version of Mosaic. It would look slightly different if it ran under the X Window (Unix) version. The actual syntax is the same.

A Sample Personal Home Page

Figure 5-1 showed a Personal Home Page whose URL is

```
file:///c¦/windows/home.htm
```

This HTML file, **home.htm**, is found in the **c:\windows** directory. Choose File¦Document Source to see the source code listed in a moment.

X
The corresponding URL for the X Window version is

```
file://localhost/usr/users/joe/home.html
```

The source code for the Home Page follows:

```
<TITLE>Joseph Radin - Home Page</TITLE>
<H1><IMG SRC="FASCINAT.GIF">Joseph Radin</H1>
<ADDRESS> Internet  : jradin@hookup.net</ADDRESS>
```

```
<ADDRESS> Phone   : ************</ADDRESS>
<ADDRESS> FAX   : ************</ADDRESS>
<ul>
<LI><A HREF="file:///d|/mosdoc/wmostoc.htm">
        MS Windows Mosaic Documentation</A>
<p>
<LI><A HREF="http://www.ncsa.uicu.edu/SDG/Software/Mosaic/Docs/UserGuide
              /Xmosaic.0.html">X Window Mosaic Documentation</A>
<p>
<LI><A HREF="http://jradin.ott.hookup.net MS Windows Server</A>
<p>
<LI><A HREF="http://www.eit.com/web/www.servers
        /www.servers.html">WWW Servers</A>
<p>
<LI><A HREF="http://wings.buffalo.edu/world">
        WorldWide WWW Info</A>
<p>
</ul>
<IMG SRC="BRIDGE.GIF">
<h1>Other Information</H1>
<ul>
<p>
<li><a href="http://www.ncsa.uiuc.edu/SDG
        /Software/Mosaic/Docs/whats-new.html">
        News with Mosaic
<li><a href="http://info.cern.ch/hypertext
        /DataSources/bySubject/Overview.html">
        CERN Virtual Library
<li><a href="gopher://wx.atmos.uiuc.edu:70/00
       /Regional/Travelers%20Forecast%20Table%20%2810%29">
       Travelers Forecast</a> for the Major Cities
<li><a href="gopher://wx.atmos.uiuc.edu/11/Canada">
        Canadian Weather Forecast</a>
</ul>
<p>
<IMG SRC="DESERT.GIF">
<p>
```

Note that the program contains text such as *Joseph Radin - Home Page,* and tags enclosed in angle brackets, as in <TITLE>. In brief, the text designates the document contents and the tags indicate how to interpret or represent the text. Use any ASCII text editor to update values as required, such as changing the title.

Chapter 6 demonstrates an automated tool that enables you to edit html files more easily and more rapidly than with an ASCII text editor.

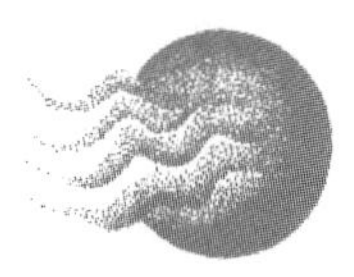

NOTE
Tags often appear in pairs, in which case the second (closing) tag includes a forward slash (/), as in </TITLE>.

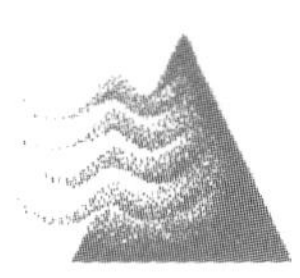

TIP
HTML does not distinguish between lowercase and uppercase letters in brackets.

Now let's examine the code line by line, starting with the top.

```
<TITLE>Joseph Radin - Home Page</TITLE>
```

Every HTML document should have a title—brief text that identifies the document. Here the title is *Joseph Radin - Home Page*. The title appears in the title bar of the displayed document.

Next comes the line:

```
<H1><IMG SRC="FASCINAT.GIF">Joseph Radin</H1>
```

This line defines the document header, displayed immediately after the box containing the document URL, as shown in Figure 5-1. This header includes a gif image and the words *Joseph Radin*. The image displayed in the figure is stored in the file **fascinat.gif**. To change the image, simply replace this file name with the name of another existing image file, making sure to specify its path.

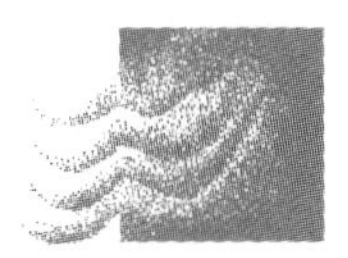

CAUTION
Mosaic won't search your directories for the image, but only looks where you tell it to look.

The tag H1 indicates that this is a level-1 header, to be formatted in a particular font. The **mosaic.ini** file, described in detail in Chapter 7, shows how to specify fonts for different header levels and other elements of your HTML documents.

The three lines:

```
<ADDRESS> Internet  : jradin@hookup.net</ADDRESS>
<ADDRESS> Phone  : ************</ADDRESS>
<ADDRESS> FAX  : ************</ADDRESS>
```

are all part of your address and are displayed in your Home Page in italics. The first address line specifies your Internet address so that people can communicate with you by e-mail. You can use the second address line to list your phone number. The third address line provides additional information such as a FAX number. In this particular instance, Joseph has hidden his phone and FAX numbers to avoid excessive calls.

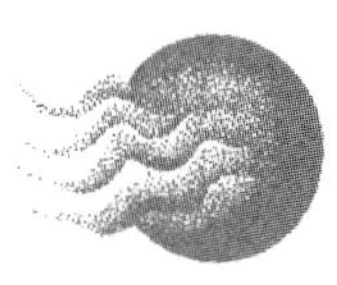

NOTE
You are not restricted to three ADDRESS lines.

The tag <p> or <P> indicates the end of a paragraph. The tag <ul> is a toggle; all text following the first occurrence is bulleted, and the tag </ul> turns bulleting off. A quick look at Figure 5-1 confirms this.

The code

```
<LI><A HREF="file:///d¦/mosdoc/wmostoc.htm">
      MS Windows Mosaic Documentation</A>
<p>
```

indicates that an underlined line will be displayed on the screen, namely MS Windows Mosaic Documentation. As you know by now, underlined text is not just static information; it is a hyperlink. Click on this hyperlink to access the list of servers. Let's look at the Home Page to see how HTML documents code hyperlinks.

The <A HREF= indicates that a URL will follow. This URL specifies the linked document, in this case

```
file:///d¦/mosdoc/wmostoc.htm.
```

The final part,

```
MS Windows Mosaic Documentation</A>
```

specifies the text displayed, identifying the linked document. As you can see, the text *MS Windows Mosaic Documentation* is underlined in Figure 5-1.

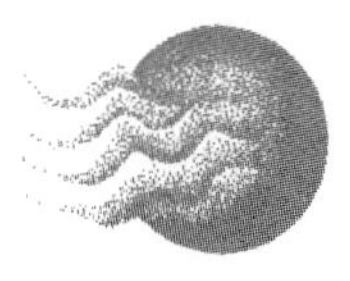

NOTE
Remember how hyperlinks work: The user clicks on the displayed text and Mosaic accesses the linked document either from a local source (see below) or a remote source (where the URL starts with a value such as http, gopher, or ftp.)

WIN
The URL for a local source is composed as follows:

```
file:///local_drive_name|/dir_name/file_name
```

X
The URL for a local source is composed as follows:

```
file://localhost/dir_name/file_name
```

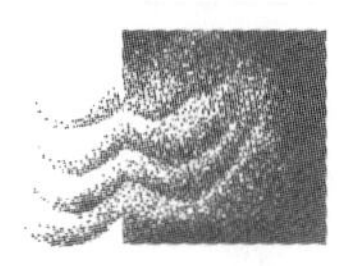

CAUTION
Get the URL syntax right, including brackets and quotation marks. If you have not made the network connection and try to link to a remote document, you will generate errors.

The next four paragraphs define the remaining four hyperlinks appearing in Figure 5-1.

```
<LI><A HREF="http://www.ncsa.uicu.edu/SDG/Software/Mosaic/Docs/UserGuide
            /Xmosaic.0.html">X Window Mosaic Documentation</A>
<p>
<LI><A HREF="http://jradin.ott.hookup.net MS Windows Server</A>
```

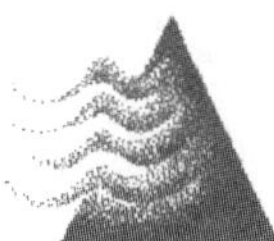

TIP
The MS-Windows Server is discussed in Appendix C.

```
<p>
<LI><A HREF="http://www.eit.com/web/www.servers
       /www.servers.html">WWW Servers</A>
<p>
<LI><A HREF="http://wings.buffalo.edu/world">
       WorldWide WWW Info</A>
<p>
</ul>
```

The line

```
<IMG SRC="BRIDGE.GIF">
```

places the image of a bridge at the bottom of Figure 5-1. The remaining lines in this personal Home Page would come into view if you scrolled down through the document shown in Figure 5-1. They place a line of text (*Other Information*) and then four more hyperlinks before a final image.

```
<h1>Other Information</H1>
<ul>
<p>
<li><a href="http://www.ncsa.uiuc.edu/SDG
       /Software/Mosaic/Docs/whats-new.html">
       News with Mosaic
<li><a href="http://info.cern.ch/hypertext
       /DataSources/bySubject/Overview.html">
       CERN Virtual Library
<li><a href="gopher://wx.atmos.uiuc.edu:70/00
      /Regional/Travelers%20Forecast%20Table%20%2810%29">
      Travelers Forecast</a> for the Major Cities
<li><a href="gopher://wx.atmos.uiuc.edu/11/Canada">
       Canadian Weather Forecast</a>
</ul>
<p>
<IMG SRC="DESERT.GIF">
<p>
```

The two final hyperlinks access Gopher servers via a Mosaic Home Page. This chapter concludes with an examination of Gopher, a major menu-based system for accessing the Internet. Remember, if you need Gopher files, you can always access them via Mosaic.

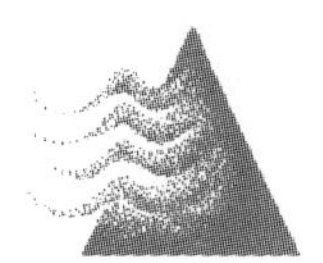

TIP
You should now be able to create your own simple Home Page. You don't have to start from scratch. You can load other Home Pages to your local disk (from the Options menu) and then cut and paste to create a personal Home Page that meets your needs. Chapter 6 explains how to automate this process even more.

Corporate and Educational Home Pages

Corporate and educational Home Pages are created and maintained at the server level of the client-server system. They tend to be long and complex. Don't let the term Home Page fool you; these Home Pages may be dozens of screens long. Because of their complexity, they may be divided into segments, each of which may be created and maintained by a different person. For example, a company might assign Home Page responsibility based upon product lines: One person handles Home Page information for product lines A and B, while another person is responsible for product line C. These people need not use the same computer; in fact they may not even work in the same city.

When a given Home Page is created and maintained on different computers, it may be called a *virtual Home Page*. Users may not even realize that a virtual Home Page is actually composed of physically different segments. Of course, if several people share Home Page responsibilities, someone must coordinate the Home Page creation and maintenance for maximum efficiency.

While corporate and educational Home Pages are created and maintained on servers, they are accessed by thousands of users from client computers. Because so many people use them and the information they contain, errors in these Home Pages tend to be critical. To some extent, a company's reputation is at stake.

A Sample Corporate Home Page

This section illustrates a corporate Home Page for the Digital Equipment Corporation (DEC), a major computer manufacturer that has decisively committed to supporting Mosaic. Figure 5-2 shows the initial screen for this Home Page, and Figure 5-3 shows the corresponding HTML source code for a portion of this Home Page.

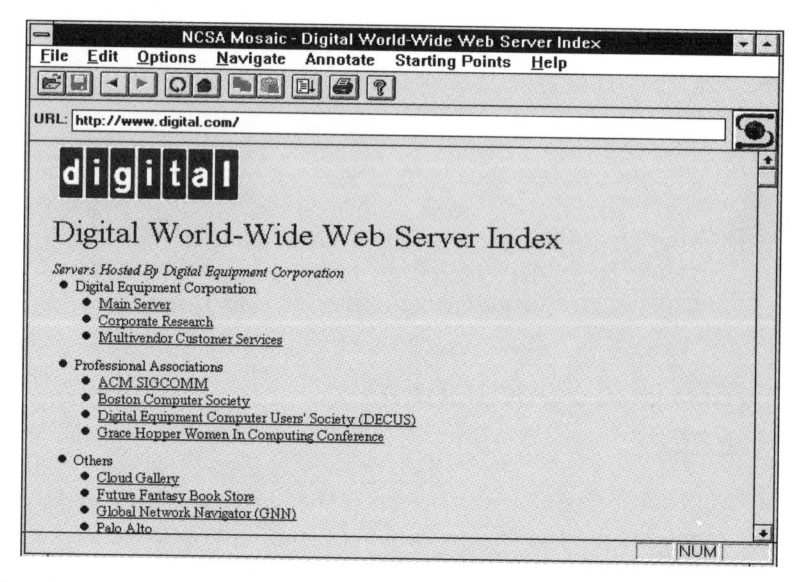

FIGURE 5-2. *The DEC Home Page*

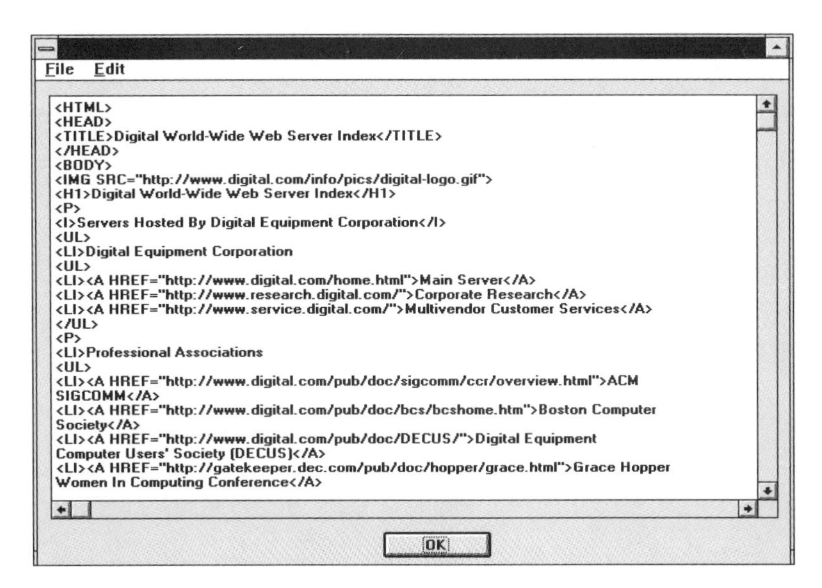

```
<HTML>
<HEAD>
<TITLE>Digital World-Wide Web Server Index</TITLE>
</HEAD>
<BODY>
<IMG SRC="http://www.digital.com/info/pics/digital-logo.gif">
<H1>Digital World-Wide Web Server Index</H1>
<P>
<I>Servers Hosted By Digital Equipment Corporation</I>
<UL>
<LI>Digital Equipment Corporation
<UL>
<LI><A HREF="http://www.digital.com/home.html">Main Server</A>
<LI><A HREF="http://www.research.digital.com/">Corporate Research</A>
<LI><A HREF="http://www.service.digital.com/">Multivendor Customer Services</A>
</UL>
<P>
<LI>Professional Associations
<UL>
<LI><A HREF="http://www.digital.com/pub/doc/sigcomm/ccr/overview.html">ACM
SIGCOMM</A>
<LI><A HREF="http://www.digital.com/pub/doc/bcs/bcshome.htm">Boston Computer
Society</A>
<LI><A HREF="http://www.digital.com/pub/doc/DECUS/">Digital Equipment
Computer Users' Society (DECUS)</A>
<LI><A HREF="http://gatekeeper.dec.com/pub/doc/hopper/grace.html">Grace Hopper
Women In Computing Conference</A>
```

FIGURE 5-3. *Source code for the DEC Home Page*

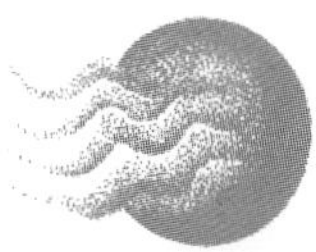

NOTE

All Home Pages are written in HTML and so can be understood by anyone familiar with this language.

CAUTION

The data gleaned from the Internet is live. Next time you access the document, the data may be updated or even removed.

Figure 5-2 is entitled Digital World Wide Web Server Index. These servers are divided into three categories: Servers hosted by DEC itself, servers connected with professional associations such as the Boston Computer Society, and others, which tend to be "fun" applications.

If you click on the Main Server hyperlink, you obtain various hyperlinks including those related to ordering DEC products, newsgroups and mailing lists, and the Alpha computer system. If you click on the Corporate Research hyperlink, you obtain a set of hyperlinks that includes corporate research publications. Instead, click on the last hyperlink in the DEC servers to obtain the Multivendor Customer Services screen shown in Figure 5-4. You can explore further from this screen.

FIGURE 5-4. *Multivendor Customer Services*

You use the second segment of the DEC Home Page to access servers belonging to professional associations, including the Boston Computer Society, the Association for Computing Machinery, and the Grace Hopper Celebration of Women in Computing.

Return to the DEC Home Page shown in Figure 5-2 by clicking on the Back icon (left-pointing arrow) in the toolbar. Then scroll to the bottom of the document and click on the please submit a problem report hyperlink to generate the Performance Trouble Report shown in Figure 5-5. Both dissatisfied and satisfied users can complete this report and send it back to DEC. User feedback helps to improve DEC's services to users.

A Sample Educational Home Page

This section illustrates an educational Home Page for the University of Illinois at Urbana-Champaign, the home of the National Center for Supercomputing Applications, where Mosaic was initially developed. The first part of this Home Page is shown in Figure 5-6. A quick look at this figure shows a wide variety of hyperlinks. Significantly the first one is Help! on Mosaic, Internet, WWW, and HTML. The second hyperlink is News and Announcements including WWW jobs.

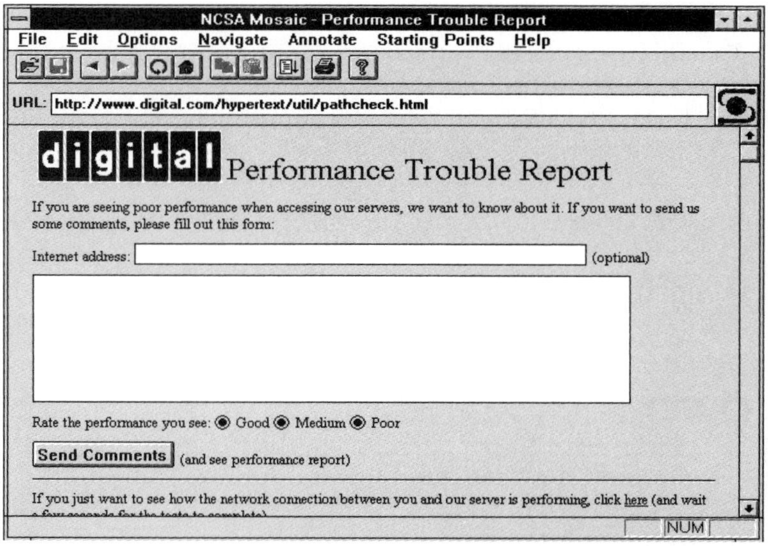

NCSA Mosaic - Performance Trouble Report

File Edit Options Navigate Annotate Starting Points Help

URL: http://www.digital.com/hypertext/util/pathcheck.html

digital Performance Trouble Report

If you are seeing poor performance when accessing our servers, we want to know about it. If you want to send us some comments, please fill out this form:

Internet address: (optional)

Rate the performance you see: Good Medium Poor

Send Comments (and see performance report)

If you just want to see how the network connection between you and our server is performing, click here (and wait

NUM

FIGURE 5-5. *A two-directional Home Page*

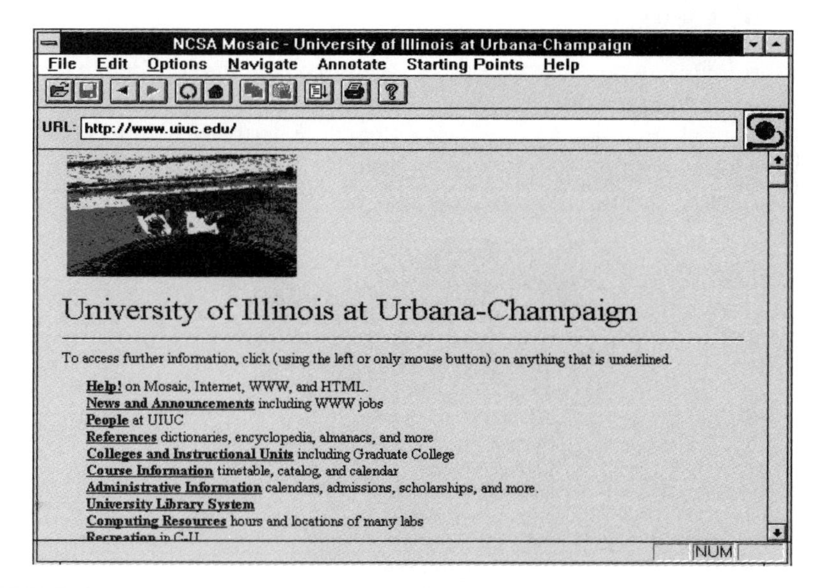

FIGURE 5-6. *University of Illinois at Urbana-Champaign Home Page*

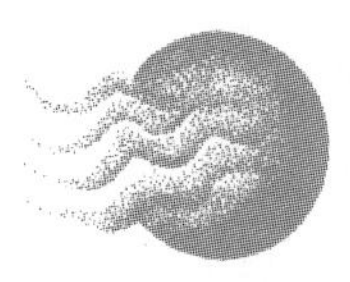

NOTE
This chapter does not show you how to create a corporate or educational Home Page, which is a relatively complex process. Chapter 6 illustrates this process by explaining how to use an automated Home Page creation and modification tool.

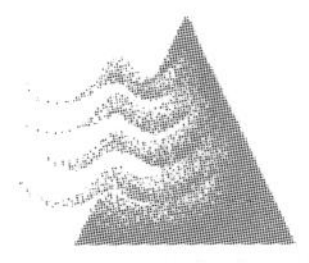

TIP
Creating a personal Home Page may only take a few hours, or perhaps a day. Depending on its magnitude and complexity, creating the initial version of a corporate or educational Home Page may take several weeks. These Home Pages must be updated constantly.

Searching Techniques

Chapter 4 illustrated several basic techniques for extracting information from the Internet via Mosaic. This chapter reviews these techniques and presents additional ones. It demonstrates two other products to search the Internet, namely Gopher and WAIS. You'll see that these techniques are interrelated; for example, if you find an interesting URL on a Gopher server you can add it to your Mosaic Hotlist.

Starting Points

The Starting Points menu, shown here, lets you specify where to begin a search. Starting points are a static list, predefined by the developers of Mosaic. They are contained in the **mosaic.ini** file as downloaded with the MS-Windows version of Mosaic or in the executable **xmosaic** file associated with the Unix version of Mosaic.

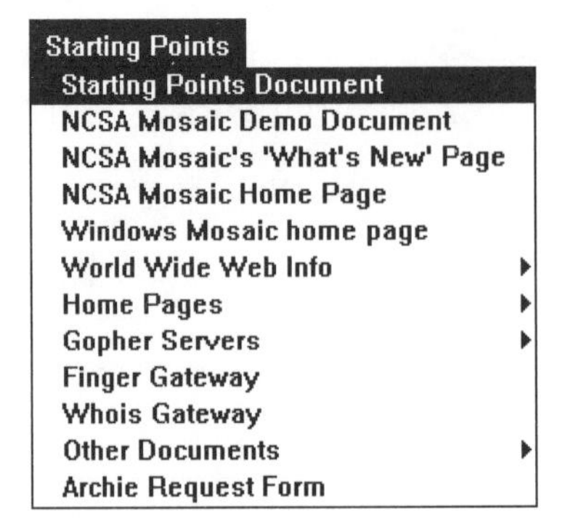

WIN
Access the starting points by clicking on the Starting Points menu item in the main menu.

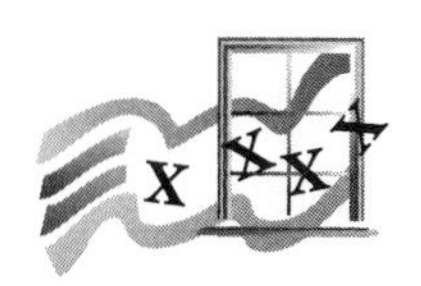

X
Access the starting points by clicking on the Internet Starting Points menu item in the Navigate menu.

Hotlists

A Hotlist is a dynamic list of your most useful URLs. To generate a Hotlist in the MS-Windows version of Mosaic, click on File¦Open URL. This generates a dialog box in which you can enter either the document URL (upper-left corner) or the document title (upper-right corner). Mosaic fills in the other value.

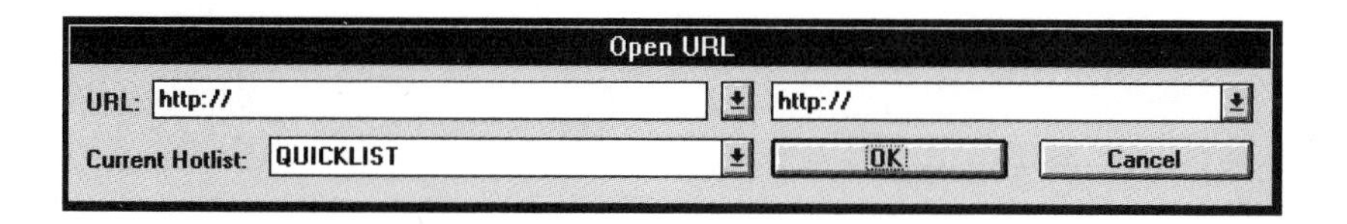

To access an item from the Hotlist, select the item by URL or document title and confirm by clicking on the OK button in the dialog box.

Initially, the Hotlist can be composed of your Starting Points menu items. Then, when you find a document of interest, you can add it to the Hotlist using a process illustrated in Chapter 7.

The History Feature

Have you ever wanted to return to a previous document, not the one you just left (which you can access by clicking on the Back button in the toolbar) but one further back? To do this, you don't have to press the Back button until you are blue in the face. Retrieving an old document is only a few menu selections away by accessing Mosaic's History feature.

WIN
Select Navigate from the main menu and then click on History to display a screen similar to the one shown in Figure 5-7. Select the desired document URL and confirm by clicking on the Load button to display that document. If you're not sure of the URL, you may have to experiment until you get the desired document. After examining the ensuing document, you can add the document to your Hotlist by clicking on the Add Current to Hotlist selection in the Navigate menu.

X

Select Navigate from the main menu and then click on Window History to display the **.mosaic-global-history** file, whose function is similar to that of the MS-Windows History screen shown in Figure 5-7. Selected lines of this file are shown here. These lines include the document URL and the document access date and time.

```
http://english-server.hss.cmu.edu/Govt.html Wed Aug  3 16:52:02 1994
http://www.sco.com/ Wed Aug  3 16:52:02 1994
http://www.llnl.gov/llnl/general.html Wed Aug  3 16:52:02 1994
http://www.sgi.com/images/newsnap.gif Wed Aug  3 16:52:02 1994
```

Once you have found the desired URL, add it to your Hotlist by accessing the Navigate/Add Current to Hotlist menu item or the Navigate/Hotlist menu item and then clicking on the Add Current button.

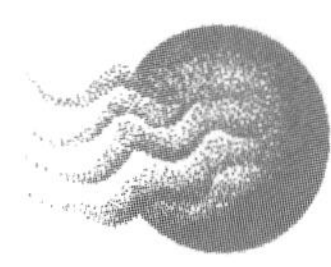

NOTE

Working with the history file is similar to working with DOSKEY; in both cases you have easy access to previously entered character strings.

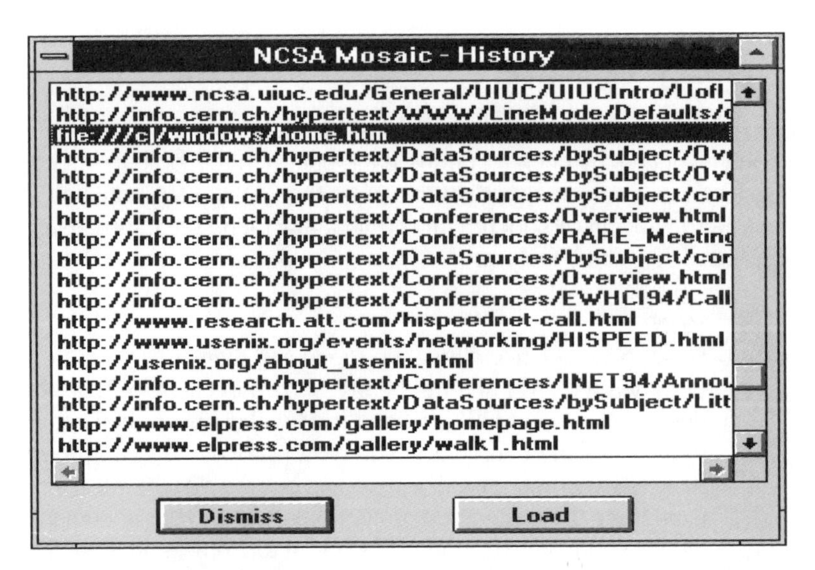

FIGURE 5-7. *Mosaic's History feature*

Gopher Servers

Chapter 1 introduced the menu-driven Gopher system. In a sense, Gopher is the World Wide Web's most important competitor for accessing the Internet. However, from Mosaic you have full access to Gopher servers around the world, in addition to the World Wide Web servers.

Like the World Wide Web, Gopher is based on the client-server mcdel. A Gopher client addresses a Gopher server to access information. There are literally thousands of Gopher servers, organized by subject. A typical search can involve multiple servers. As a client, you don't have to worry about where the information is located. You ask for it, and Gopher gets it. When you find a menu you like, you can tell Gopher to save it as a *bookmark.* Then the next time you need it, you can find it fast. Bookmarks are similar to Mosaic Hotlists.

Applying the Search Techniques

This section applies two types of search techniques. In the first type of search, you know where to look. Under these circumstances you use what are called *structured search techniques*; you begin with the personal Home Page. In the second type of search, you don't know where to look. In this case you use *unstructured search techniques*; you begin with the Starting Points and apply Gopher and WAIS systems to search and explore the Internet. You should master both techniques.

Structured Search Techniques

You start structured searches either by clicking on a specific hyperlink or by entering the corresponding URL manually.

Don't enter the URL manually more than once. Either put it in your Hotlist or use the editor and put the URL in your Home Page as discussed earlier in this chapter. For example, enter a line such as the following:

```
<li><a href="http://info.cern.ch/hypertext
       /DataSources/bySubject/Overview.html">
       CERN Virtual Library
```

to place the Virtual Library in your Home Page. After doing so, click on the Home icon in the toolbar to reload your Home Page for the changes to take effect.

Scroll down the Home Page and then click on the CERN Virtual Library hyperlink to generate the WWW Virtual Library screen.

The WWW Virtual Library is an interesting catalog, whose contents differ from the list of Commercial World Wide Web Servers used in Chapter 4. Which one should you use? Both. They say you can't be too rich or too thin. Provided that you know how to control yourself, you also can't have too many informative Internet addresses. Appendix A provides a wealth of useful Internet addresses.

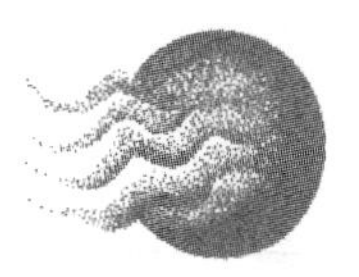

NOTE
The WWW Virtual Library was created using the WAIS system discussed at the end of this chapter.

This section applies structured search techniques to find tourist information such as restaurant guides and transit schedules. As any well-organized tourist knows, a successful trip starts with planning from home, in this case from our Home Page.

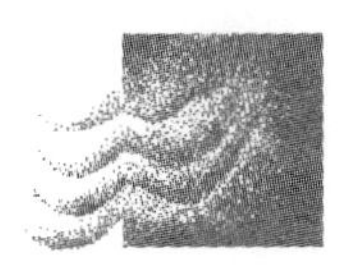

CAUTION
Some of the following screens contain large images. Be prepared to wait. Who wants a tour book without illustrations anyway?

The next step is to click on the World Wide Web Info hyperlink. What if you don't have this hyperlink? Access the File menu and select the Open URL menu item. In the URL text box enter the following value:

```
http://wings.buffalo.edu/world
```

and confirm by clicking on the OK button.

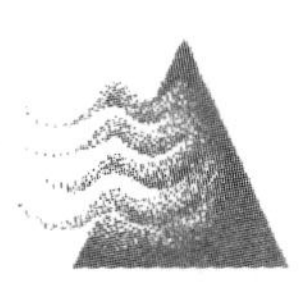

TIP
Add this URL to your Hotlist using the process described earlier in this chapter.

Clicking on this hyperlink displays a map of the world. We don't have time to search the whole world now, so specify North America by clicking on the rightmost rectangle in the map to generate the screen shown in Figure 5-8.

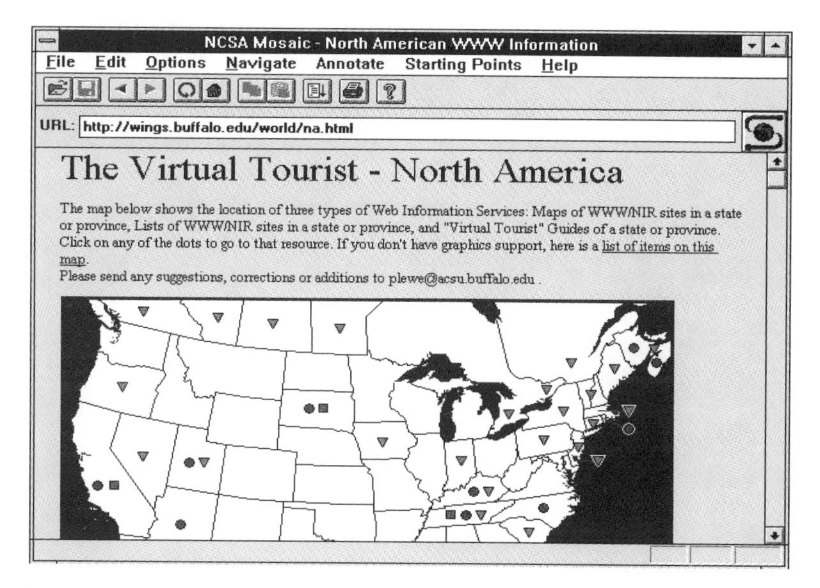

FIGURE 5-8. *The Virtual Tourist sees North America*

This figure contains three types of information for American states and Canadian provinces:

- Maps
- Lists of special sites
- Virtual Tourist guides

Click on the map of California to generate the California World Wide Web Servers document. Scroll down this screen until you see the Bay Area Restaurant Guide hyperlink and click on it to generate the screen shown in Figure 5-9. Since the guide includes over 12,000 restaurants, you probably won't want to just scroll aimlessly down the list. Instead you must choose one of the following search strategies:

- You can search by restaurant name.
- You can search by city name.
- You can search by type of food.
- You can determine the final search strategy yourself.

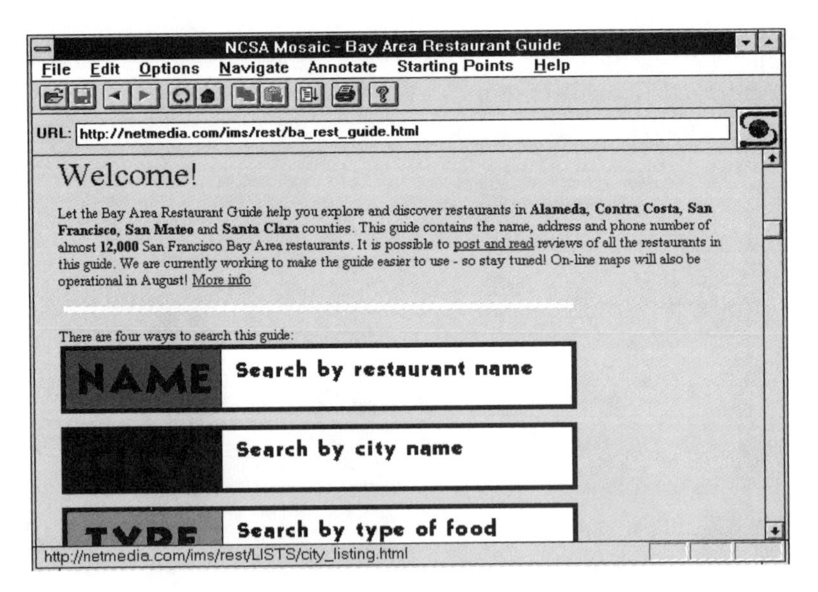

FIGURE 5-9. *Looking for the right restaurant*

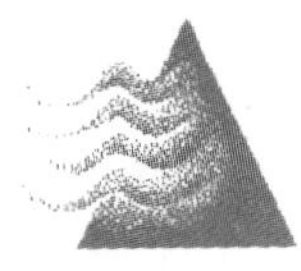

TIP
Add interesting URLs to your Hotlist so you don't have to repeat this search process the next time you're hungry in the Bay Area.

Click on the Search by city name hyperlink and then scroll down the resultant screen until you get to Oakland. Would you believe that the guide lists 868 restaurants for Oakland? Interestingly enough, none of them was reviewed for this Restaurant Guide.

One key aspect in choosing a restaurant, especially in an unfamiliar city, is its location. Let's take a short ride on the world famous Bay Area Rapid Transit (BART). Return to the map of California screen, scroll down it, and click on the Bay Area Transit Information hyperlink. Step through a few clearly marked screens, specifying first "BART" and then "system map" to generate the screen shown in Figure 5-10. BART provides a similar routing map that lets you find the suggested route between the originating and destination stations.

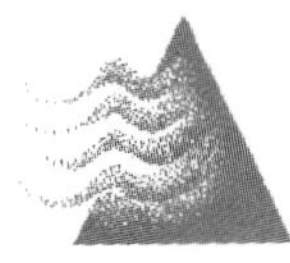

TIP
By using the Hotlist, the History feature, or the Back and Forward buttons, you can easily switch back and forth between the restaurant pages and the map showing you how to get there. Try finding a restaurant that painlessly, armed only with a restaurant guide and a roll of quarters for the phone!

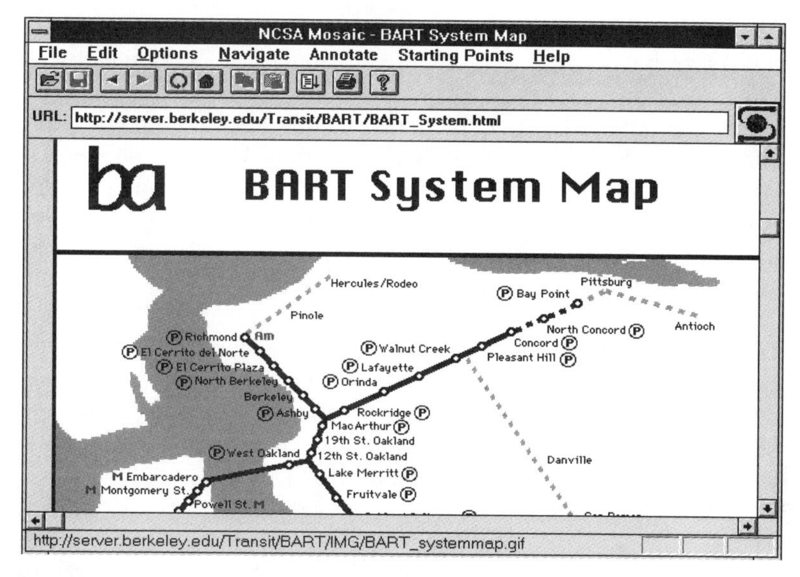

FIGURE 5-10. *The Bay Area Rapid Transit System Map*

Unstructured Search Techniques

Use structured search techniques if you know where you are going. Use unstructured search techniques if you don't. The following unstructured search techniques are explained in this section:

- Using Veronica to search Gopher menus for a specified character string.
- Using WAIS to find a specified character string within the Internet documents themselves.
- Making use of the news to find newsgroup articles whose name contains the specified character string.

Use Veronica, WAIS, or the news to begin searching. Once you get the Gopher menus, Internet documents, ftp servers and directories, or newsgroup, you usually continue searching for the exact reference. As you can see, the search process is not explicitly defined; that's why this type of search is called an unstructured search.

Veronica

Veronica is a search tool that looks for Gopher menus containing selected key words in document titles. For example, with a simple Veronica command you can find all Gopher menus that include the phrase *Comic Books*. Then you can select these menu items as desired, using Gopher commands. Veronica allows you to use Boolean operators such as "and," "or," and "not." For example, you could ask for a menu item including *Comic Books* or *Illustrateds*. It takes some time to learn how to specify commands to avoid two common problems with Veronica searches: finding too few responses and finding too many responses.

As mentioned, unstructured searches begin from the Starting Points menu option. The following exercise looks for references that contain the character string *Open Computing*. In this case, you don't really know where the references are. However, during the search you will learn about available sources. This additional information may lead to subsequent structured searches.

To begin a search using Veronica, click on the Starting Points menu and select the Starting Points Document menu item to generate the screen shown in Figure 5-11.

Click on the Veronica hyperlink to generate a list of Veronica servers that perform key word searches. From the ensuing screen, select uni-koeln (The University of Cologne in Germany) and enter **Open Computing** under Search Index, as shown in Figure 5-12.

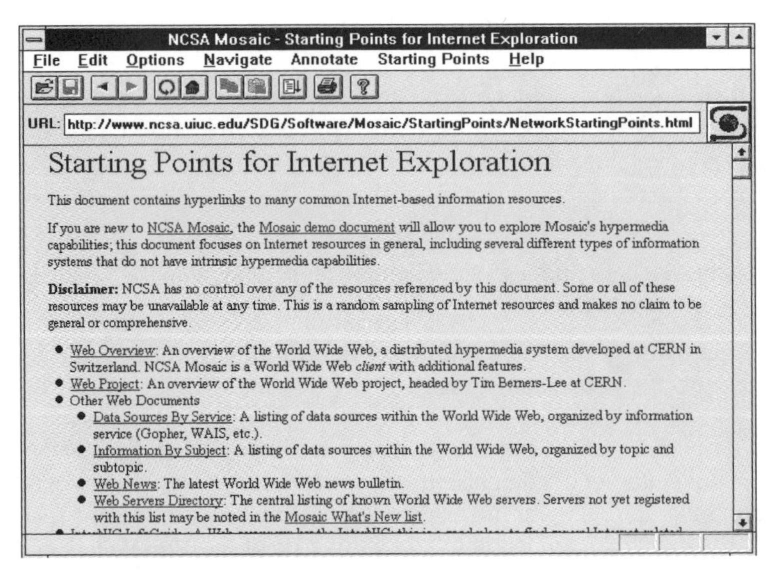

FIGURE 5-11. *Starting Points Document*

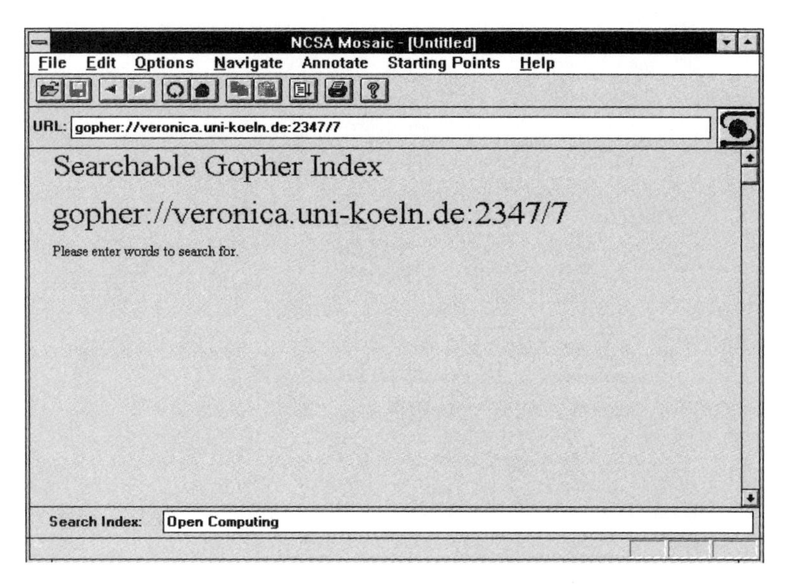

FIGURE 5-12. *A Veronica key word search*

Press the ENTER key to confirm the Search Index, which generates a list like the one shown in Figure 5-13. (The complete list of references is several screens long.) Then click on the third item from the top to generate the book reference shown in Figure 5-14.

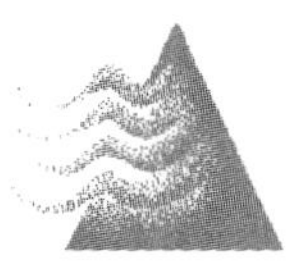

TIP
Add the URLs shown either in Figure 5-13 or 5-14 to your Hotlist if you plan to access them again for the Search Index *Open Computing.*

WAIS

WAIS, Wide Area Information System, is an Internet service originated by three firms: Apple Computer, the makers of the Macintosh; Dow Jones, an information service; and Thinking Machines, a manufacturer of computer servers. The goal of WAIS is to provide on demand a list of sources that contain references to the key words you supply. Unlike Veronica, which searches document titles, WAIS does a *full-text search.* It looks for the key word within the articles themselves. In other words, if an article contains the character string *Perfume* but the document title does not, WAIS will find it but Veronica will not.

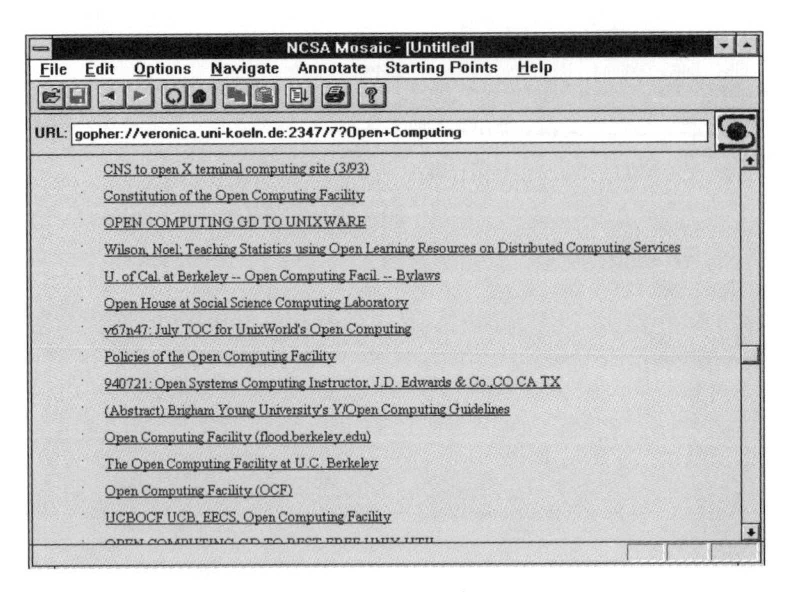

FIGURE 5-13. *References to open computing*

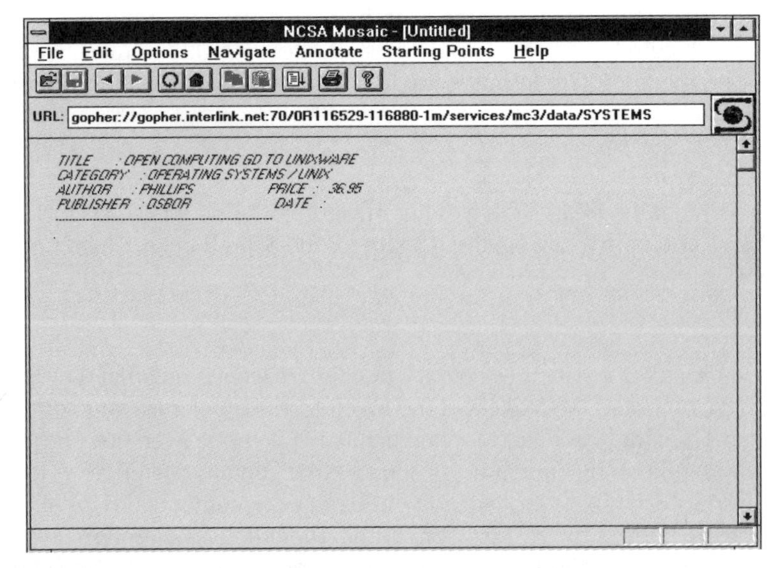

FIGURE 5-14. *An Open Computing book*

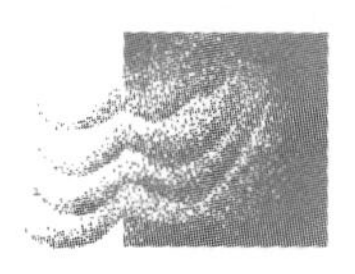

CAUTION
WAIS's power can make it difficult to use. Improperly specifying your search string leads to either too many or too few items appearing in the list of sources.

Return to the the Starting Points Document, scroll down, and select the WAIS Starting Point hyperlink. In the ensuing screen, click on the hyperlink associated with the CERN WAIS interface to generate the list of data sources shown in Figure 5-15. Click on the hyperlink Directory of Servers at the top of this list. Enter the Search Index **www and mosaic** and press the ENTER key to confirm. This generates the screen shown in Figure 5-16, listing the articles in order according to the frequency with which the designated words were found. Click on a hyperlink to display desired articles or to save them in a file.

Accessing the News

It is hard to overestimate the importance of news on the Internet. Arguably, for a majority of Internet users, news is the major, if not the only, Internet application. Mosaic provides full, easy access to Internet news.

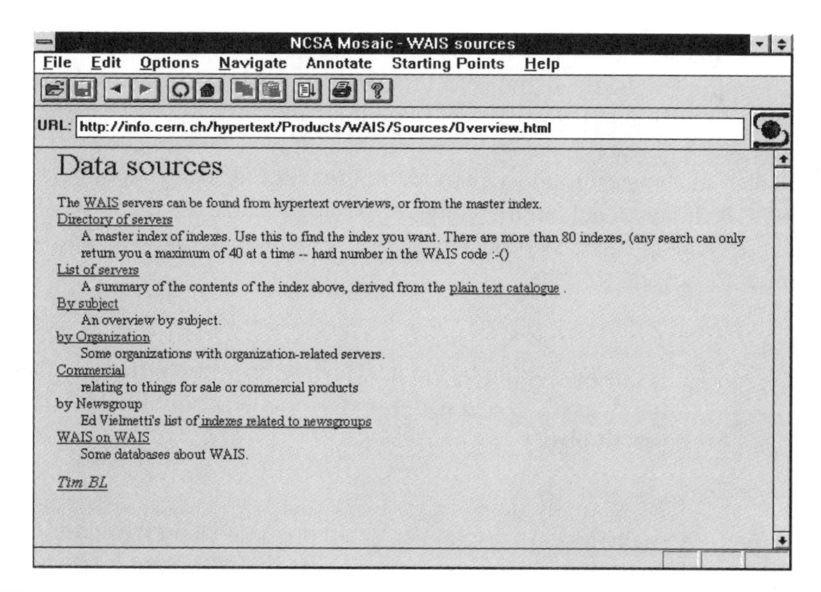

FIGURE 5-15. *A list of WAIS data sources*

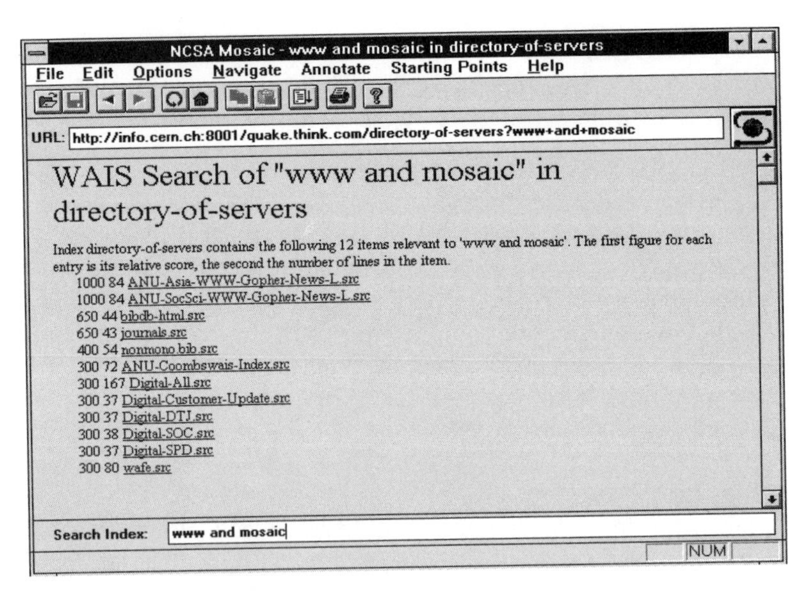

FIGURE 5-16. *Specifying a WAIS search*

Mosaic makes it very easy to get the Internet news, as you'll learn here. Click on the Starting Points menu and select the Starting Points Document. Move down the screen to the Usenet Newsgroups hyperlink, and click on it to generate the screen shown in Figure 5-17. Then click on the alt hyperlink to generate a screen listing alternative newsgroups, which are not monitored. This output is shown in Figure 5-18. Finally, click on the alt.bbs hyperlink to access a list of articles in the alternative bulletin board systems newsgroup, as shown in Figure 5-19. The list continues for several screens.

TIP
If necessary, you may want to review how newsgroup names are formed, as discussed earlier in the chapter. Here the URL is **news:alt.bbs**.

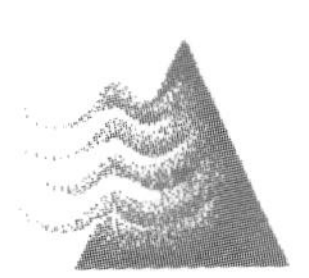

TIP
Usually Mosaic displays selected documents on the screen. However, you can also transfer documents to your local disk by using the following procedure for your system.

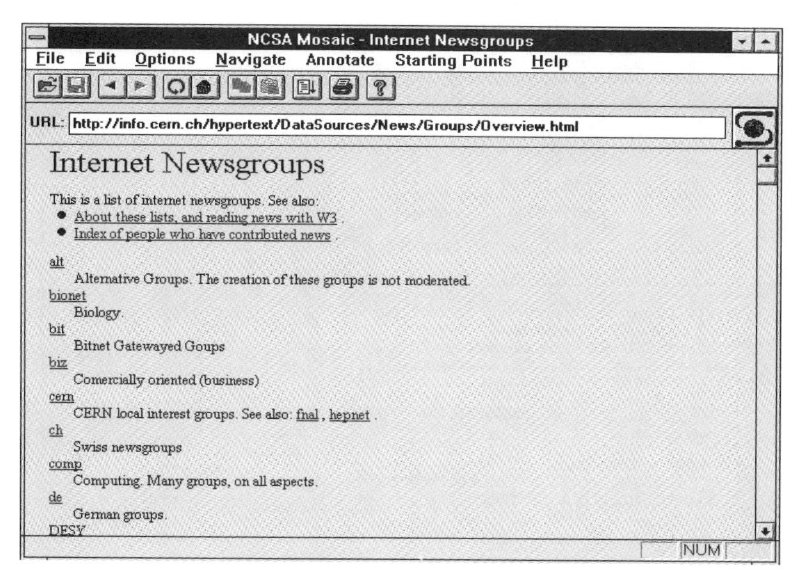

FIGURE 5-17. *Internet newsgroups*

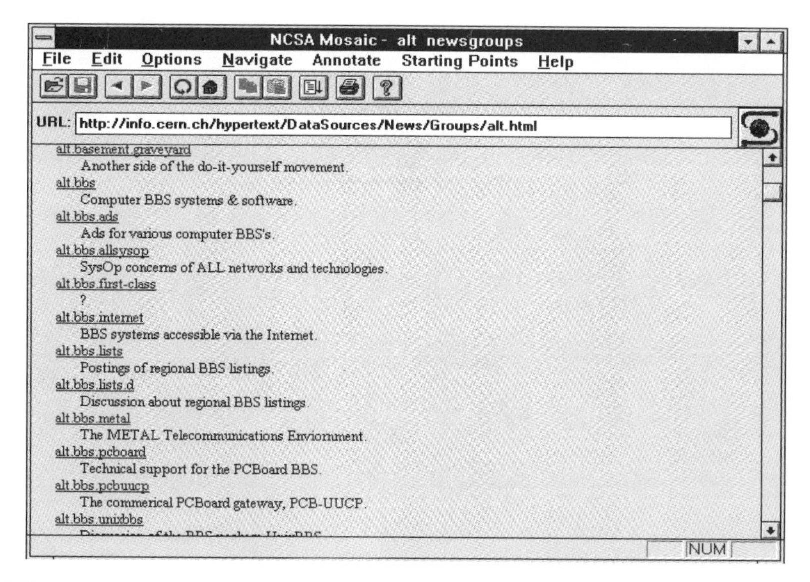

FIGURE 5-18. *Alternative Internet newsgroups*

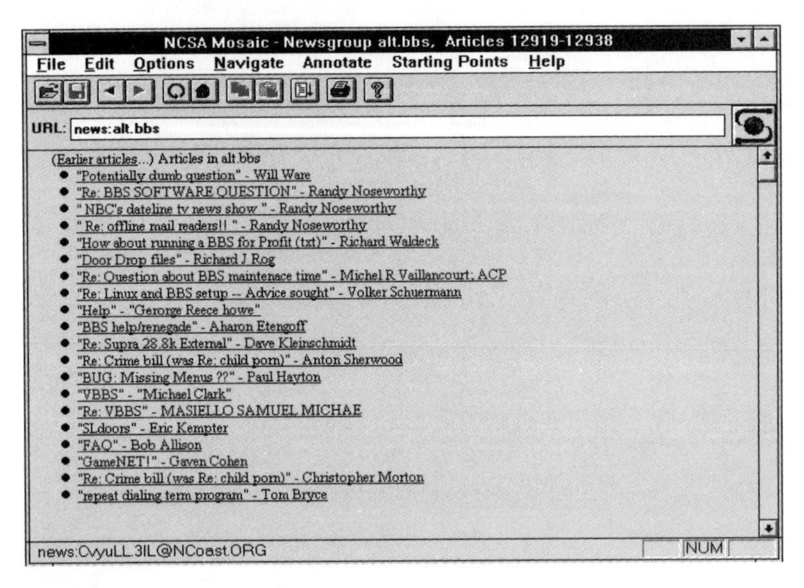

FIGURE 5-19. *News on alt.bbs*

The Hypertext Markup Language (HTML)

Chapter 5 presented a short example of HTML, Hypertext Markup Language, used for creating documents that can be stored on the World Wide Web and browsed via Mosaic. HTML is the format that determines how Home Pages and other documents on the Web will appear. This chapter extends your mastery of HTML documents in two ways. First it explains the HTML elements in detail. Then it introduces a software package called HoTMetaL that you can use to automate the process of creating and modifying HTML documents.

HTML Elements

An HTML document is composed of *elements,* which in turn are made up of entities. *Entities* represent graphical characters with special meanings in the markup. *Markup* consists of special codes that format the document. The markup consists primarily of elements. Entities include the less than sign (<), the greater than sign (>), the *&* (to denote an ampersand), and the *"* (to denote the double quote sign). They also include standard foreign language character sets.

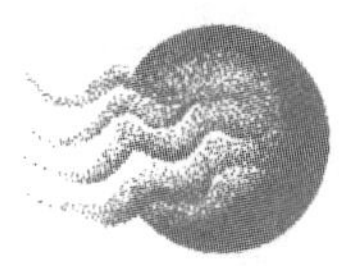

NOTE
Elements may be enclosed within other elements.

Some elements describe the entire document or link it to other documents. Other elements format portions of the text. Still others are used to insert graphics within the document. Every HTML element starts with a tag which appears within angle brackets (<>). The most important HTML elements are discussed in the sections that follow. Figure 6-1 shows the HTML source code associated with the NCSA Home Page, part of whose onscreen image is shown in Figure 6-2.

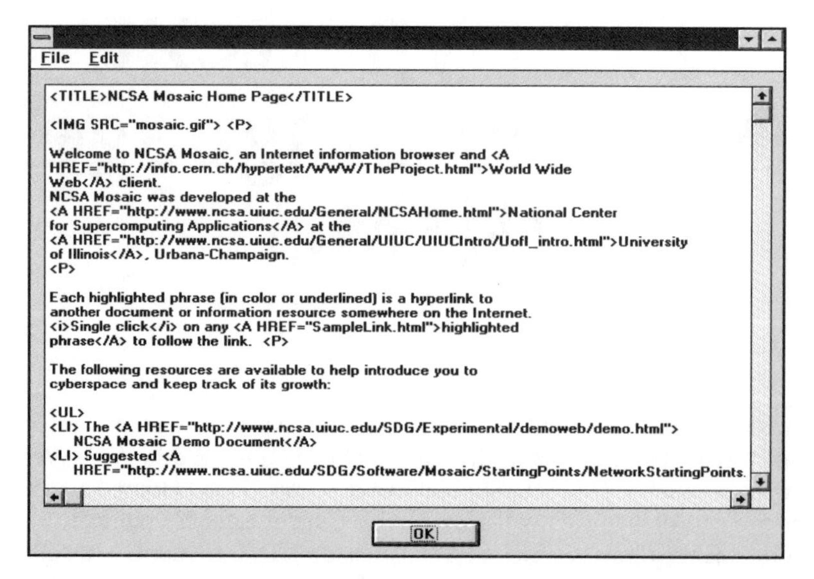

```
<TITLE>NCSA Mosaic Home Page</TITLE>

<IMG SRC="mosaic.gif"> <P>

Welcome to NCSA Mosaic, an Internet information browser and <A
HREF="http://info.cern.ch/hypertext/WWW/TheProject.html">World Wide
Web</A> client.
NCSA Mosaic was developed at the
<A HREF="http://www.ncsa.uiuc.edu/General/NCSAHome.html">National Center
for Supercomputing Applications</A> at the
<A HREF="http://www.ncsa.uiuc.edu/General/UIUC/UIUCIntro/UofI_intro.html">University
of Illinois</A>, Urbana-Champaign.
<P>

Each highlighted phrase (in color or underlined) is a hyperlink to
another document or information resource somewhere on the Internet.
<i>Single click</i> on any <A HREF="SampleLink.html">highlighted
phrase</A> to follow the link.  <P>

The following resources are available to help introduce you to
cyberspace and keep track of its growth:

<UL>
<LI> The <A HREF="http://www.ncsa.uiuc.edu/SDG/Experimental/demoweb/demo.html">
    NCSA Mosaic Demo Document</A>
<LI> Suggested <A
    HREF="http://www.ncsa.uiuc.edu/SDG/Software/Mosaic/StartingPoints/NetworkStartingPoints.
```

FIGURE 6-1. *HTML source code for NCSA Mosaic Home Page*

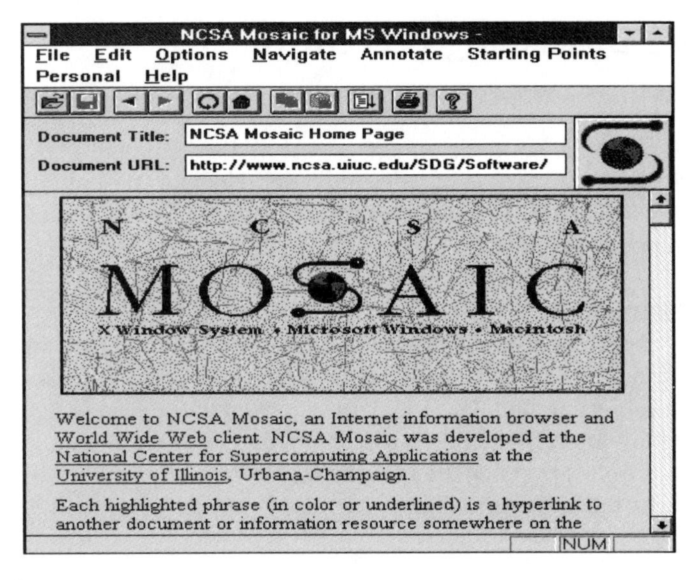

FIGURE 6-2. *NCSA Mosaic Home Page*

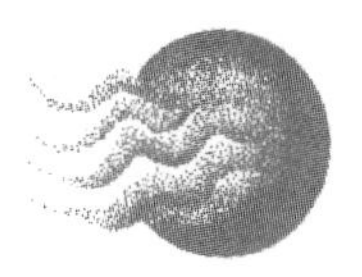

NOTE
Tags often appear in pairs. The first denotes the beginning of the tagged item and the second, which includes a slash, denotes the end of the tagged item. For example, in Figure 6-1 <TITLE>NCSA Mosaic Home Page</TITLE> denotes the document title "NCSA Mosaic Home Page" as it appears in Figure 6-2. Tags are not case sensitive; <TITLE> and <Title> have the same effect.

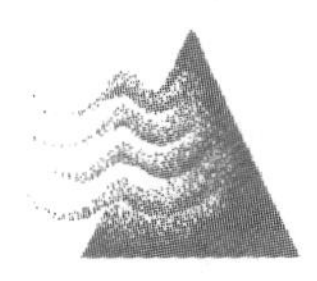

TIP
To view such code under Mosaic, choose Document Source from the File menu (see Chapters 2, 3, and 8).

HTML Structure

Any HTML document includes a HEAD element and a BODY element. The HEAD element contains all information describing the document as a whole, such as the document title. The BODY element follows the HEAD element and contains all information that is part of the document, such as document text and character highlighting.

Elements Describing the Entire Document

Elements describing the entire document appear in the document's HEAD element. You can code these elements in any desired order. Elements describing the entire document include the TITLE element and other elements, whose discussion is beyond the scope of this text.

The TITLE Element

The TITLE element appears in the Mosaic history file and in the Document Title box or the Window Title bar. It should identify the document contents in a general fashion. If the subject of the title is too narrow, potential users will ignore the document. If the subject of the title is too broad, users will waste their time and network resources by accessing the document unnecessarily.

For example, the first line in Figure 6-1

```
<TITLE>NCSA Mosaic Home Page</TITLE>
```

clearly specifies the document title.

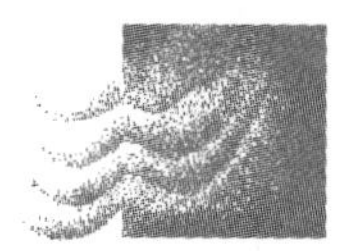

CAUTION
Limit the title's length to 64 characters.

Text Formatting Elements

Text formatting elements describe how parts of the document appear in the output. You can place them within the document's BODY element in any desired order. HTML provides the following text formatting elements: headings, anchors, paragraph marks, address style, blockquote style, lists, preformatted text, and character highlighting.

Headings

HTML enables you to enter up to six levels of headings by using the tags <h1> to <h6>. The highest level heading is <h1>, which is recommended for the start of a hypertext document. Standard practice discourages jumping more than one level of heading—for example, going directly from <h2> to <h4>.Typically, headings are represented as shown in Table 6-1.

HEADING	DESCRIPTION
<h1>	Produces a heading that is boldface, centered, in a very large font, followed by one or two blank lines, and always printed at the top of a new page.
<h2>	Produces a heading that is boldface, left justified, in a large font, and both preceded and followed by one or two blank lines.
<h3>	Produces a heading that is italic, slightly indented from the left margin, in a large font, and both preceded and followed by one or two blank lines.
<h4>	Produces a heading that is boldface, indented from the left margin more than h3, in a normal font, and both preceded and followed by one blank line.
<h5>	Produces a heading that is italic, indented from the left margin the same as h4, in a normal font, and preceded by one blank line.
<h6>	Produces a heading that is boldface, indented from the left margin, in a normal font, and preceded by one blank line.

TABLE 6-1. *HTML Heading Levels*

Anchors

An *anchor* element is a text segment marking the beginning and/or end of a hypertext link. The following statement is an example of an anchor:

```
<A HREF="http://info.cern.ch/hypertext/WWW/TheProject.html">
World Wide Web</A>
```

The HREF characters indicate that the World Wide Web is hypertext and refers to the beginning of a link. Click on the hypertext and you will access the document whose URL is **http://info.cern.ch/hypertext/WWW/TheProject.html**.

Paragraph Marks

Paragraphs are indicated by the <P> tag. Most other tags occur in pairs, but the <P> tag does not. It is used to separate two blocks of text that would otherwise appear together, as shown in several examples in Figures 6-1 and 6-2. Typically, <P> generates a line or a half line of space between paragraphs, except for Address or Preformatted text elements. Sometimes the <P> tag generates a small left indent on the first line of normal text.

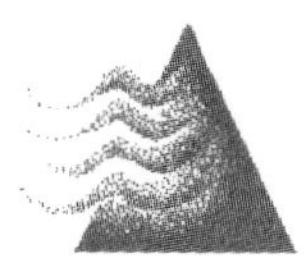

TIP
You should not use the paragraph tag to generate blank space around headings, lists, or other elements. Headings generate their own space.

Address Elements

The Address element is often used to indicate the city or e-mail address, signature, and authorship associated with a document. It typically appears at the top or the bottom of a document. It is usually in italics and/or right justified. The Address element automatically includes a paragraph break. Following is a sample address:

```
<ADDRESS>Internet jradin@hookup.net</ADDRESS>
```

Blockquote Tags

The <blockquote> and </blockquote> tags appear in pairs. They denote the beginning and the end of text quoted from another source, such as:

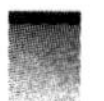

```
<blockquote>To be or not to be, that is the
question...</blockquote>
```

Typically, such text is slightly indented from the left and right margins and/or italicized. A paragraph break occurs and one or more blank lines precede and follow the quoted text. Sometimes a vertical line of ">" characters appears in the left margin, as in Internet quotations.

List Tags

A list is a series of paragraphs, each of which may be numbered or preceded by a special mark such as a bullet. The tags <UL> and </UL> denote the beginning and end, respectively, of a list whose elements are preceded by bullets. The tags <OL> and </OL> denote the beginning and end, respectively, of a list whose elements are numbered. The single tag <LI> denotes the beginning of a list element.

You can see the beginning of a bulleted list at the bottom of Figure 6-1.

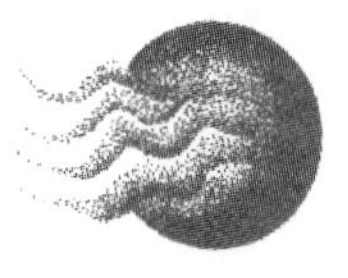

NOTE
The elements in a list may be simple text or complex elements such as links and anchors, as shown in this example.

Preformatted Text Tags

A pair of <PRE> tags denotes preformatted text such as a computer listing. This text is displayed in a monospaced font, one in which all characters have the same width.

Character Highlighting

Character highlighting provides emphasis for sections of text, such as hyperlinks. Character highlighting tags appear in pairs. Three commonly used character highlighting tags are <B> for boldfacing, <I> for italics (or a slanted font if italics are not available), and <U> for underlined text. The tags <CITE> and </CITE> denote a citation, which typically appears in italics.

Figure 6-1 includes a simple example of character highlighting, <i>Single click</i>, which places this phrase in italics on the screen.

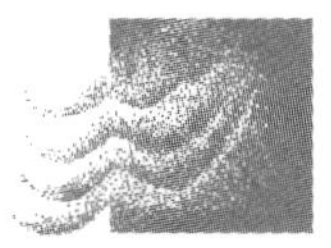

CAUTION

In general, you can use anchor elements and character highlighting such as italics in preformatted text. Sometimes, however, the character highlighting may not be rendered correctly. You cannot use elements such as Headings and Addresses that define paragraph formatting in preformatted text.

Comment Tags

A comment is an explanation or another nonexecutable statement. Comments do not directly affect Mosaic, but make it easier to understand and modify the HTML document. They are enclosed within <!— and —>. For example, the following line is a valid comment:

```
<!— HTML coding is easy —>
```

but the following line is an invalid comment:

```
<— HTML coding is fun !—>.
```

Embedded Images

You can also use tags and elements to insert embedded images into an HTML document.

The IMG Element

The IMG element represents a graphical image and enables you to insert an icon or a small graphical image within a document. This element includes the mandatory SRC element denoting the URL of the document to be embedded, and the optional ALIGN element describing the physical relationship between the embedded graphic and the surrounding text. The ALIGN element may assume the value TOP, MIDDLE, or BOTTOM.

The second line of Figure 6-1 includes

```
<IMG SRC="mosaic.gif">
```

which inserts a gif image in the Home Page, as shown in Figure 6-2.

Image elements can exist within anchor elements; for example, you could have the code

```
<A  HREF="address_value"> <IMG SRC="switch.gif">CLICK HERE</A>
```

By clicking on the image or on the hypertext (CLICK HERE), you will activate the link to the document whose URL is defined by the HREF value (address_value).

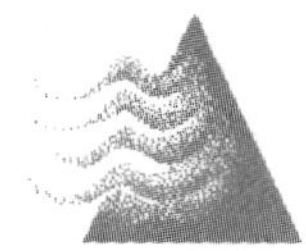

TIP
The IMG element is used for embedded images in text or in anchors.

More Information on HTML

For more information on the Hypertext Markup Language and its use, consult the following URL:

```
http://info.cern.ch/hypertext/WWW/Bibliography.html
```

The ensuing screen is entitled Bibliography for the World Wide Web. In the section entitled Manuals and Primers, click on the list of manuals hyperlink, whose URL is

```
http://info.cern.ch/hypertext/WWW/Papers/PaperManuals.html
```

The first part of this document explains how to download the manuals. You want the HTML Specification Manual, whose file name is **html-spec.txt** (text) or **html-spec.ps** (Postscript).

Creating HTML Documents with HoTMetaL

As described earlier in this chapter, HTML documents, like most electronic documents, contain text and markup. Markup consists of special codes such as tags that format the document. These codes direct the computer how to structure and process the document text, and how to relate the present document to other documents. While tag syntax is not complicated, it is easy to make several errors when specifying tags. HoTMetaL relieves you of the burden of dealing with tag syntax; you simply specify what you want, and the software takes care of the syntax, for example, by creating or modifying tags. Usually, you can accomplish what you want with only a few menu selections, instead of having to pour over a technical HTML manual.

HoTMetaL is an unsupported product distributed by SoftQuad Inc. SoftQuad has developed a supported version, HoTMetaL Pro, which is available for sale. It has additional functionality such as WYSIWYG editing of complex tables and spell checking.

Downloading the HoTMetaL Software

You can obtain HoTMetaL from the following locations via anonymous ftp:

```
ftp.ncsa.uiuc.edu/Mosaic/contrib/SoftQuad
ftp.ifi.uio.no:/pub/SGML/HoTMetaL
sgml1.ex.ac.uk:SoftQuad
doc.ic.ac.uk:/pub/packages/WWW/ncsa/contrib/SoftQuad
ftp.cs.concordia.ca: pub/www
ftp.cc.gatech.edu in/pub/gvu/www/pitkow/misc
ftp.sunet.se:/pub/www/Mosaic/contrib/SoftQuad
ftp.uco.es:/www
olymp.wu-wien.ac.at:/pub/sgml/exeter/SoftQuad
ftp.informatik.uni-freiburg.de:/pub/WWW/editors/HoTMetaL
gatekeeper.dec.com:/pub/net/infosys/Mosaic/contrib/SoftQuad
```

Here is how to download the product from one of the available sources, gatekeeper.dec.com, from Digital Equipment Corporation.

```
ftp gatekeeper.dec.com
login: anonymous
password: ident
```

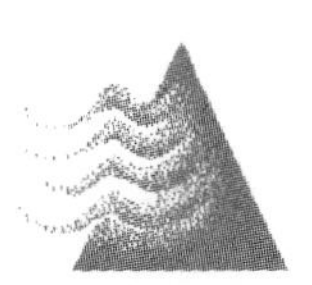

TIP
Most servers require a full e-mail address such as **user_name@mail_address** as a password, and also make you set the transfer mode to binary.

Use the following code to download documentation from the **/pub/net/infosys/Mosaic/contrib/SoftQuad/hotmetal** directory:

```
ftp>cd /pub/net/infosys/Mosaic/contrib/SoftQuad/hotmetal
ftp>get FAQ
ftp>get Changes
ftp>get README
```

WIN

Use the following code to download the HoTMetaL to a MS-Windows-based system:

```
ftp>cd MS-WINDOWS
ftp>get hotmetal.exe
ftp>get readme.txt
```

X

Use the following code to download the HoTMetaL to a Unix-based system (Sun SPARC station):

```
ftp>cd SPARC-Motif
ftp>get sq-hotmetal-1.0b.README
ftp>get sq-hotmetal-1.0b.tar.Z
```

Installing HoTMetaL

There are versions of HoTMetaL that run on MS-Windows and on Sun SPARC stations (Unix). (In the near future, there should be versions for Macintosh, HP-UX, and Digital Unix.) The installation procedures for these two versions differ, and are described in the following sections.

Installing under MS-Windows

To install the MS-Windows version of HoTMetaL you need to have MS-Windows 3.1 as well as about 5MB of free disk space on your hard drive.

Create a directory to receive the HoTMetaL files using the following code. (The d: drive is used as an example.)

```
$ md d:\hotmetal
```

Change to the specified directory and unpack the archive by activating the self-extracting distribution file:

```
$ cd d:\hotmetal
d:hotmetal>hotmetal.exe
```

Then execute the following MS-Windows procedure to create a program group.

1. Open Windows.
2. Select the File¦New option in the Program Manager.
3. Click on the Program Group button in the dialog box and click on the OK button to confirm.
4. Enter **SoftQuad HoTMetaL** in the Description text box and click on the OK button to confirm.

To create a program item, follow these steps:

1. Open Windows and activate the HoTMetaL program group window.
2. Select the File¦New option in the Program Manager.
3. Click on the Program Item button in the dialog box and click on the OK button to confirm.
4. Enter **HoTMetaL** in the Description text box.
5. Type the following command in the Command Line text box. (The d: drive is used as an example.)

   ```
   d:\hotmetal\sqhm.exe-sqdir d:\hotmetal
   ```

6. Type **d:\hotmetal** in the Working Directory text box and click on the OK button to confirm.

This adds the HoTMetaL icon to the program group window. Now you should be able to run the HoTMetaL program by double-clicking on the icon.

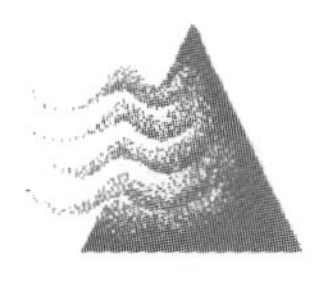

TIP
For useful information, print the manual, which is in **hotmetal/doc/hotmetal.ps**, on a Postscript printer. The manual is about 60 pages.

Installing under Unix

If you have a Sun or compatible SPARC system running SunOS 4.*x* or Solaris 1.*x* and you have about 6MB of free disk space, you can install SoftQuad HoTMetaL. You must be running Open Windows 3, or X11R4 or later. HoTMetaL has been run on Solaris 2.3 (SPARC) and may run on Solaris 2.1 or 2.2.

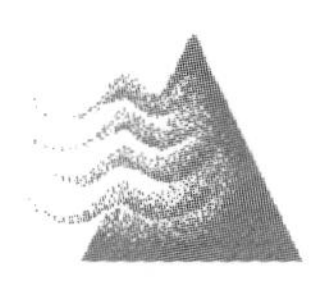

TIP
Check available disk space with the df command.

Installing HoTMetaL usually does not require root (system administrator) permission. If you have a problem in the installation, access the **readme** file for more information about local installation.

Put a copy of the configuration file in your home directory, such as **/usr/users/user_name**. You can place the HoTMetaL files in a directory of your choice—for example, **/usr/local**.

Change to the specified directory by entering a command such as

```
$ cd /usr/local
```

If you have a gzipped tar file (whose extension is **.tar.gz**) enter

```
$ gunzip < sqhotmetal-1.0.tar.gz ¦ tar xvf
```

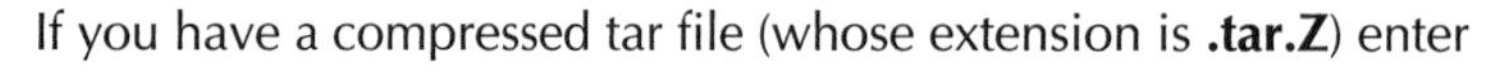

If you have a compressed tar file (whose extension is **.tar.Z**) enter

```
$ zcat sqhotmetal-1.0.tar.Z ¦ tar xvf
```

Both sh and ksh users set their path as follows:

```
$ SQDIR=/usr/local/hotmetal;
$ PATH=$SQDIR/bin:$PATH;
$ export SQDIR PATH
```

Users of csh, bash, and zsh set their path as follows:

```
$ setenv SQDIR /usr/local/hotmetal;
$ setenv PATH
$ SQDIR /bin:$PATH
```

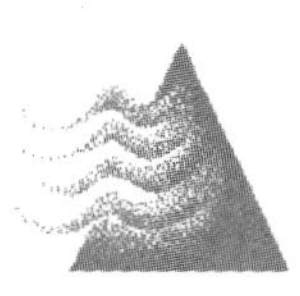

TIP
To find out your shell, type **echo $SHELL**. The response indicates your shell.

Install X app-defaults (default values for X Window system) by copying the configuration file **Sqhm** into **/usr/lib/X11/app-defaults** or merging the entries with your own **.Xdefaults** file in your **home** directory.

```
cp hotmetal/Sqhm /usr/lib/X11/app-defaults/Sqhm
```

or

```
cp hotmetal $HOME
```

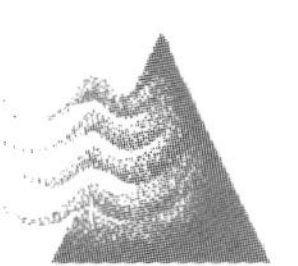

TIP
For more information on X app-defaults, see *X Window Inside and Out,* by Levi Reiss and Joseph Radin (Osborne/McGraw-Hill, 1992).

To run the HoTMetaL program simply enter

```
sqhm &
```

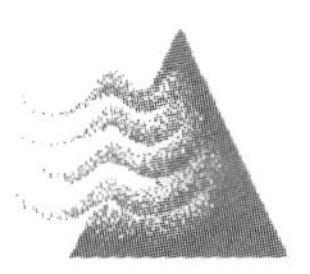

TIP
For useful information, print the manual, which is in **hotmetal/doc/hotmetal.ps**, on a Postscript printer. The manual is about 60 pages.

Use the following command to print the HoTMetaL manual:

```
lpr -Pps $SQDIR/hotmetal/doc/hotmetal.ps
```

Configuring HoTMetaL MS-Windows Version

The MS-Windows version of HoTMetaL includes the **sqhm.ini** file. You can modify this file to configure HoTMetaL, much as you might modify the **windows.ini** file to configure MS-Windows. For example, the default viewer for gif files is given in the following line in **sqhm.ini**:

```
view_gif = d:\psp\psp
```

Because we are using another viewer, as described in Appendix B, we changed this line to:

```
view_gif = d:\wingif\wingif
```

Another example is the value

```
html_browser = mosaic
```

Obviously, we do not want to change this value (see the File¦Preview option discussed in the section on HoTMetaL menus).

Overview of HoTMetaL Menus

When you first start HoTMetaL MS-Windows version, it displays a welcome screen and then opens an empty file **scratch1.htm** as shown here:

The menu bar contains the following six menu bar options:

- The File menu for processing files
- The Edit menu for editing document text
- The View menu for formatting the screen and displaying different views of the document structure
- The Markup menu for inserting and modifying markup (tags, and so forth) and assuring that the markup follows syntax rules
- The Help menu for obtaining online help
- The Windows menu for switching between open files and arranging the onscreen windows.

The File Menu

The File menu, shown here, contains options for creating, opening, saving, closing, previewing, and publishing HoTMetaL files and exiting the program. The menu options follow.

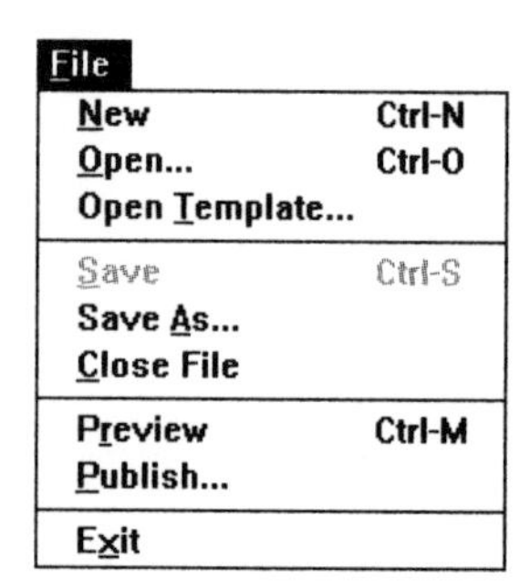

The New Option

The New option creates an empty HTML file whose name is **scratch** followed by a consecutive number starting with 1, and whose extension is **.htm**.

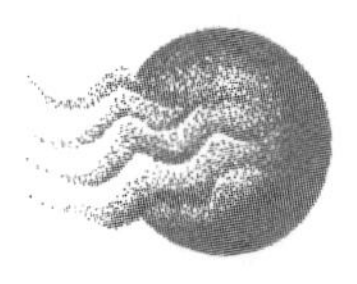

NOTE
You can change this name when saving the file.

The Open Option

The Open option opens a saved file according to information that you specify in the dialog box shown in Figure 6-3.

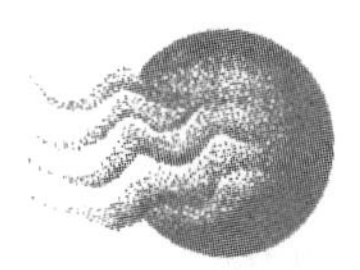

NOTE
You can open multiple files.

You can have HoTMetaL check whether a document conforms to HTML rules—for example, whether the appropriate tags appear in pairs—when it opens the file. You determine whether or not HTML checks the document in this manner by turning on the Rule Checking On option in the Markup menu.

FIGURE 6-3. *Open dialog box*

The Open Template Option

The Open Template option enables you to open a standard template and create and modify documents based upon this template. As its name indicates, a template is a predefined structure for documents. You can enter text into a template without having to enter any markup.

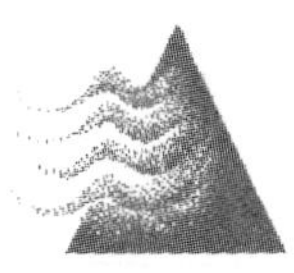

TIP

Appropriate use of this option speeds document creation and reduces error frequency. Figure 6-4 shows the Open Template dialog box.

The Save Option

The Save option saves the current document as an HTML file (.htm) by default. The text file type (.txt) is also available.

The Save As Option

The Save As option saves a copy of the current document under a different name after confirmation. The user provides the new name, the new drive, and directory designation, if applicable.

The Close File Option

The Close File option closes the current file without exiting HoTMetaL.

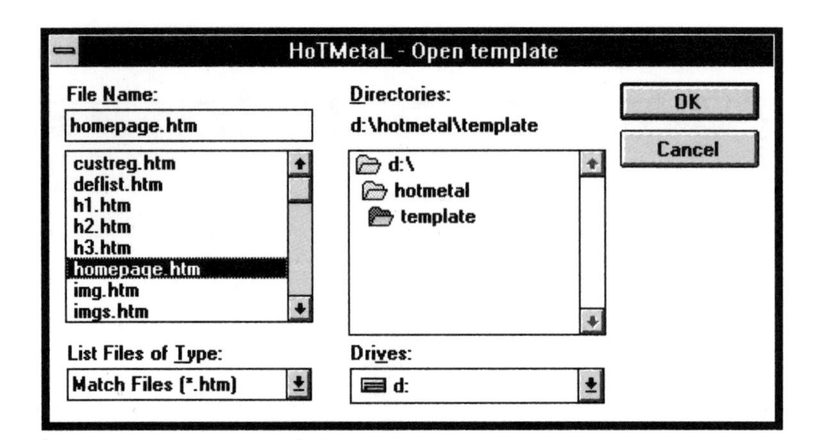

FIGURE 6-4. *Open template dialog box*

The Preview Option

The Preview option is a major HoTMetaL feature. This option actually launches Mosaic, enabling you to preview your document, as shown in Figure 6-5. All Mosaic options discussed in Chapters 2, 3, and 8 are available. Make your Internet connection before activating this option so that you can activate hyperlinks to remote documents. Otherwise, you will be only in the local mode and any attempt to access remote documents will generate error messages. Exiting the Preview option returns you to the HoTMetaL program.

TIP
Use the Preview option frequently to see the actual effects of your changes.

The Publish Option

The Publish option changes the URL from a value specifying local access such as

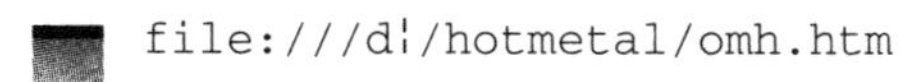

```
file:///d¦/hotmetal/omh.htm
```

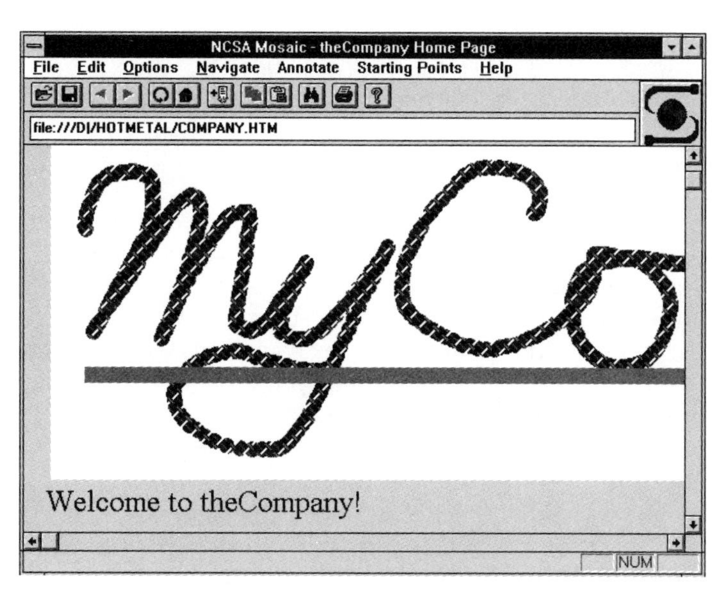

FIGURE 6-5. *Previewing your document via Mosaic*

to a value such as

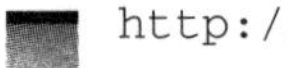

```
http://server_name/omh.htm
```

making the file available on the Internet.

The Exit Option

As you can guess from its name, the Exit option lets you quit HoTMetaL. The system will ask you for confirmation.

The Edit Menu

The Edit menu, shown here, contains commands for editing documents, including undoing previous commands, cutting and pasting from the document to the Clipboard, finding, and replacing. The menu options follow.

Edit	
Undo	Ctrl-Z
Redo	
Cut	Ctrl-X
Copy	Ctrl-C
Paste	Ctrl-V
Find and Replace...	Ctrl-F
Find Next	F3

The Undo Option

The Undo option reverses commands; you can use this command when you've entered a command in error. In general, you can undo most commands. However, you cannot undo File menu commands, text selection, or scrolling and windowing commands. You also cannot reverse an Undo command with another Undo command. Furthermore, you cannot undo any commands you entered prior to saving the file. The number of commands that you can undo is specified in the *undo_limit* configuration variable, whose default value is 10.

The Redo Option

The Redo option reverses the effect of an Undo command.

The Cut Option

The Cut option erases the selected text from the current document and places it in the Clipboard so that you can paste it into another document or into a different location in the same document.

The Copy Option

The Copy option copies selected text from the current document and places it in the Clipboard so that you can paste it into another document or into a different location in the same document.

The Paste Option

The Paste option transfers the contents of the Clipboard into the specified document. You use the Paste option after performing either a Cut or Copy operation.

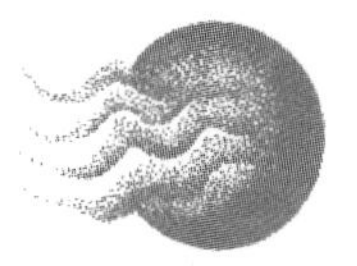

NOTE

The contents of the Clipboard remain the same until you cut or copy something new to the Clipboard, enabling you to paste the same thing repeatedly.

The Find and Replace Option

The Find and Replace option locates and replaces strings within the document. This command leads to a dialog box that provides several choices. For example, you can automatically replace all found occurrences or review each occurrence individually, replacing it or moving on to the next occurrence. You can specify wildcards, such as "*" and "?", in the search patterns and search for tags. You can search forward or backward, conduct case-sensitive or non-case-sensitive searches, and locate entire words only or both entire words and embedded strings.

Figure 6-6 shows the automatic replacement of the string UnixWare by the string Mosaic. The Wrap option causes HoTMetaL to search the entire document. After the replacement, HoTMetaL displays a confirmation window. If you try to exit the document without saving your changes, you will see the two buttons Discard Changes (exit anyway) and Cancel (return to the document). The Replace All button replaces all occurrences of the search string without requesting confirmation.

The Find Next Option

The Find Next option is a shortcut for repeating the previously specified search. It performs a search for the search string specified in the previous Find & Replace dialog box. Once a search string has been specified, this command has the same effect as clicking the Find button in the Find & Replace dialog box.

HoTMetaL - Find & Replace

Find: UnixWare

Replace: Mosaic

Find In:

☐ Whole Words ☐ Case Sensitive ☒ Wrap

☐ Backwards Search ☐ Find Patterns

Find | Replace | Replace then Find | Replace All | Cancel

FIGURE 6-6. *Find & Replace dialog box*

The View Menu

The View menu, shown here, contains commands for formatting the display window of HoTMetaL documents. The menu options follow.

View

Hide Tags	Ctrl-W
Show Context Window	F12
Show Structure Window	F11
Show Image	
Character...	Ctrl-B
Separation...	
Load Styles...	

The Show/Hide Tags Option

The Show/Hide Tags option is a toggle that specifies whether tags such as <P> are visible or invisible, as in Figure 6-7, which shows hidden tags. The default value for a new document is visible tags.

The Show Context Window Option

The Show Context Window option opens the Context View window, shown here, which indicates the sequence of nested elements up to the insertion point. This window will not show the structure of the entire document, only the hierarchical context containing the current position. It also displays the sequence of open tags at the current position.

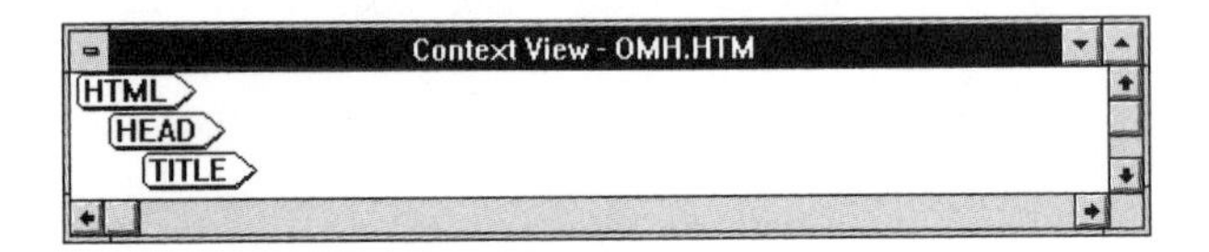

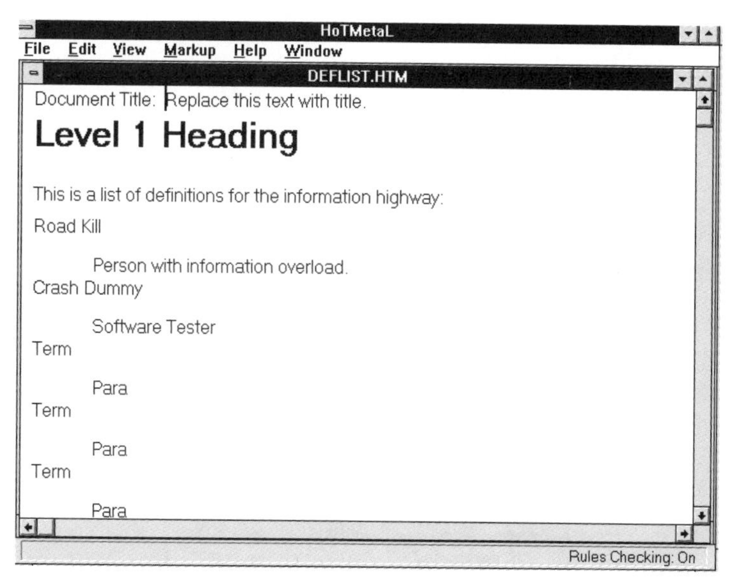

FIGURE 6-7. *Hidden tags*

The Show Structure Window Option

The Show Structure Window option opens the Structure View window. This window shows the structure of the entire document including tags as shown here:

You can edit the document via this view.

The Show Image Option

The Show Image option displays graphical images such as gif files embedded into edited documents.

WIN

Before applying the Show Image option, initialize the configuration variable view_gif in the configuration file **sqhm.ini** as described earlier. The viewer path must be set correctly according to the viewer installation procedure discussed in Appendix B.

X
Before applying this option, initialize the appropriate configuration variable such as view_gif in the configuration file **Sqhm.** The viewer (such as **xv**) path must be set correctly according to the viewer installation procedure discussed in Appendix B.

The Character Option

The Character option controls character formatting such as fonts and justification. Simply make selections in a dialog box like the one shown in Figure 6-8. Choosing the Adopt Current check box indicates that you want to use the present value.

The Separation Option

The Separation option lets you specify the vertical spacing between elements via a dialog box like the one shown here:

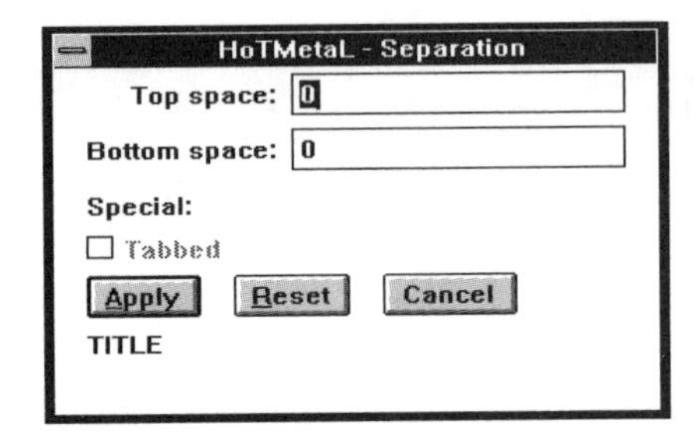

The Load Styles Option

The Load Styles option enables you to load styles (formatting information) via a dialog box like the one shown in Figure 6-9. Click on a desired style to change the document's format.

The Markup Menu

The Markup menu, shown here, contains commands for "marking up"—or adding HTML codes to—a document. The menu options follow.

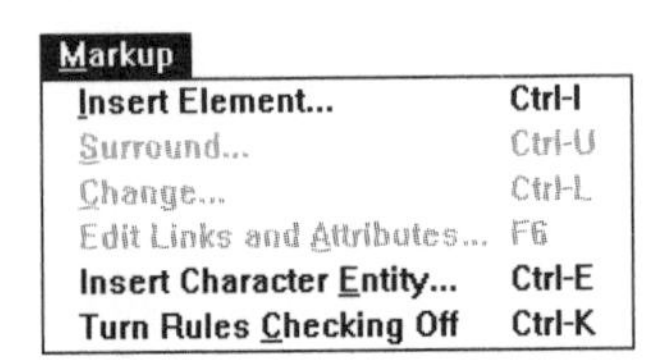

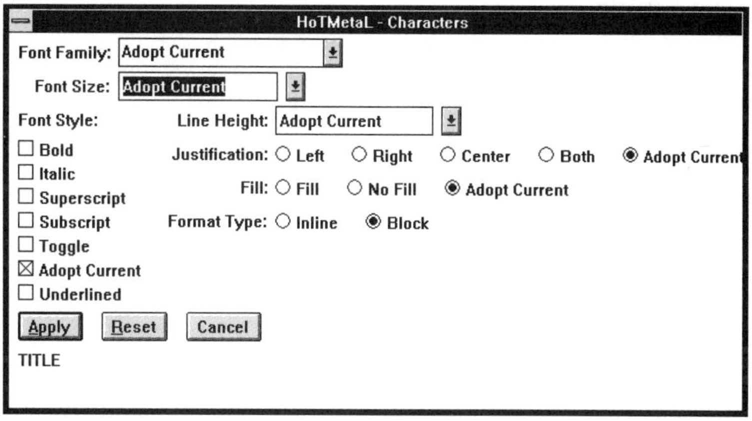

FIGURE 6-8. *Character dialog box*

The Insert Element Option

The Insert Element option creates a new element such as a document title. Selecting this option generates a dialog box with an alphabetical list of elements from which you can choose. You can also determine whether HoTMetaL checks that the inserted element conforms to the rules, as discussed at the end of this section.

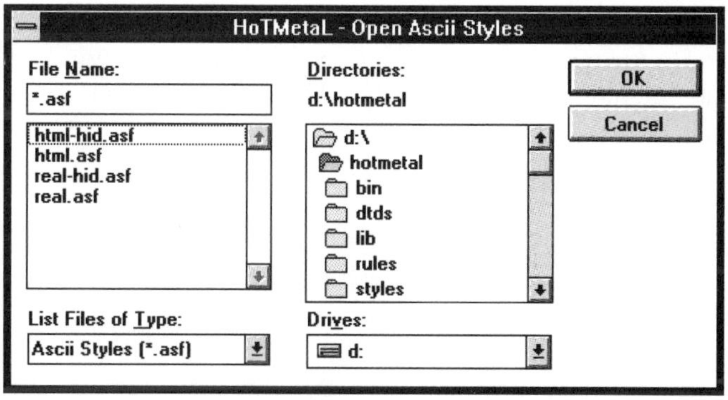

FIGURE 6-9. *Open Ascii Styles dialog box*

The Surround Option

The Surround option encloses inserted text within the HTML BODY element. Use this option before completing the document. When you invoke this option, the Surround dialog box appears with a list of elements that can surround the selection and still leave the document correctly marked up. Choose an element in the same manner as when inserting an element.

The Change Option

The Change option modifies the type of the current element. For example, you can use this option to change an element from italic to bold. This will change the way the element looks in the resultant document.

The Edit Links and Attributes Option

The Edit Links and Attributes option permits you to modify an attribute that represents the URL, enabling you to establish a link to a different document.

The Insert Character Entity Option

The Insert Character Entity option allows you to access foreign language character sets and other special characters. Refer to the discussion of entities at the beginning of this chapter.

The Turn Rules Checking Off/On Option

The Turn Rules Checking Off/On option is a toggle. When this option is on, HoTMetaL finds most markup errors, such as not including the proper ending code.

The Help Menu

The Help menu, shown here, contains options for finding out more about HoTMetaL and upgrading it to the commercial version. The options are discussed briefly below.

The About HoTMetaL Option

The About HoTMetaL option provides information about the HoTMetaL software. The screen shown in Figure 6-10 is for the free, unsupported software. The screen shown in Figure 6-11 is for HoTMetaL Pro, the commercial, fully supported version.

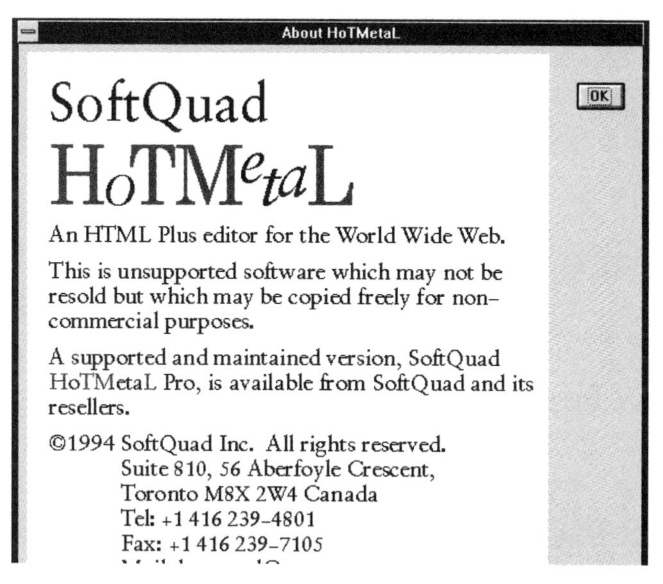

FIGURE 6-10. *About HoTMetaL*

FIGURE 6-11. *About HoTMetaL Pro*

The Window Menu

The Window menu is a standard MS-Windows menu offering options such as cascading or tiled windows. See your MS-Windows documentation for more information.

Home Page Creation

As an example, you will learn how to create a corporate Home Page (see Chapter 5) for "theCompany." The preliminary version of a Home Page is shown in Figure 6-5. The final results are shown in Figures 6-12 and 6-13.

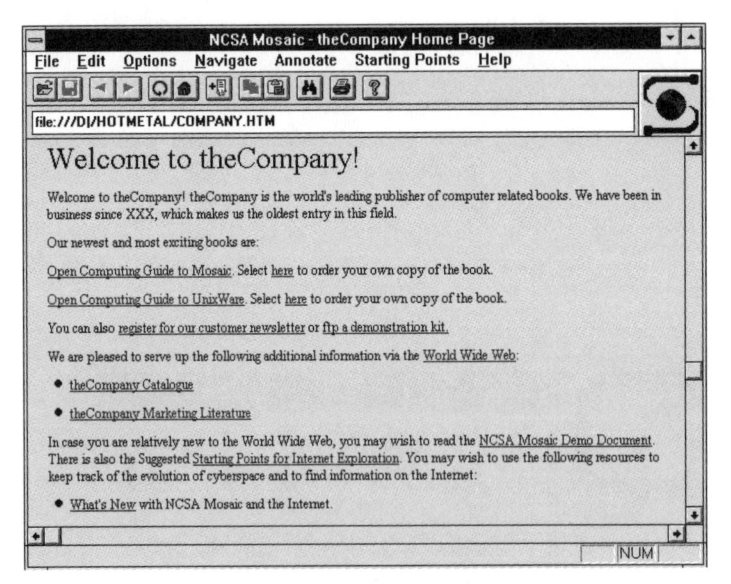

FIGURE 6-12. *First Part of theCompany Home Page*

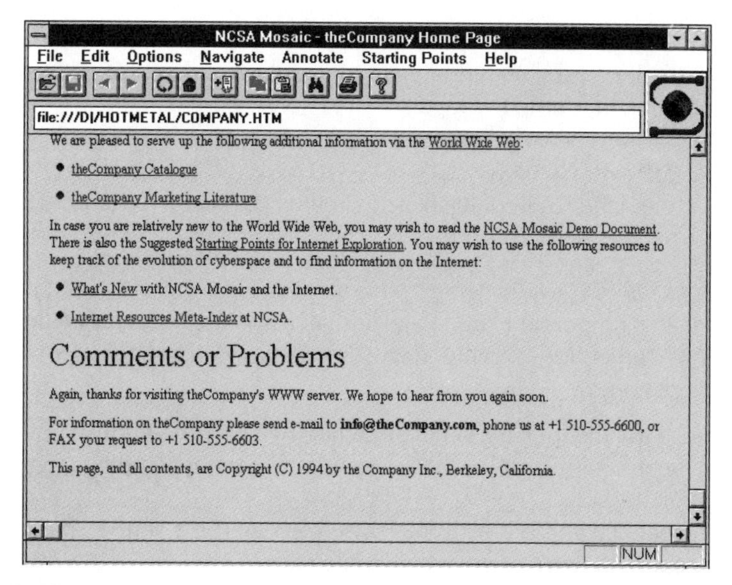

FIGURE 6-13. *Second Part of theCompany Home Page*

The steps are as follows:

1. Execute HoTMetaL and select New from the File menu. Then select the standard HOMEPAGE.HTM template and save it, for example, as OMH.HTM.
2. Edit this file by changing the text inside the tags. For example, in the document title replace MyCompanyInc. with theCompany.
3. Apply the proper settings from the Markup menu such as editing characters and the appropriate links. Replace default template links such as **product1.htm** with links associated with theCompany.

4. Insert a gif file containing theCompany logo inside the IMG tags in the upper half of the screen. For example, replace **mylogo.gif** with a gif file containing theCompany logo.

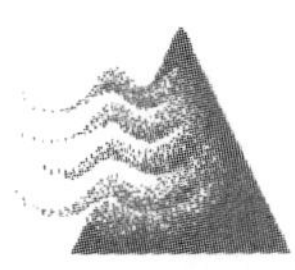

TIP
View the result of each step via the File¦Preview menu option

5. Place this corporate Home Page and all associated files on an appropriate WWW server, for example, www.thecompany.com. Servers are discussed in Appendix C.
6. Publish the file (place it on the Internet) by using the Publish option of the File menu. For example, change to http://www.thecompany.com/omh.html.

You just created a corporate Home Page. You can use a similar, but perhaps simpler, procedure to create a personal Home Page.

Customizing Mosaic

Now that you have completed an extensive tour of Mosaic and its information gathering facilities, it's time to learn how to customize Mosaic so that you can use these tools in a smooth and efficient manner. MS-Windows users have a lot of latitude in configuring Mosaic via the **mosaic.ini** file. Unix users have much fewer options for configuring Mosaic, but can still make certain changes via the **Mosaic** file, as described at the end of this chapter.

Configuring Mosaic for Windows

Chapters 4 and 5 explained how to use Mosaic to gather information. Chapter 5 presented the theory and practice of setting up your own Home Page. Now you are ready to configure Mosaic, customizing it to meet your specific needs.

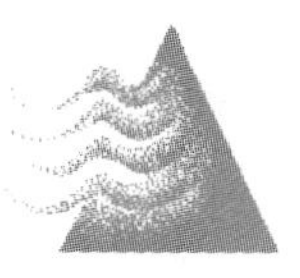

TIP
In general, use the menus (see Chapters 2, 3 and 8) for *temporary* modifications and the **mosaic.ini** file for more *permanent* changes. For instance, if you want to change a font once, or only sporadically, you can use the menus. But if you want to change the font permanently, you can modify the **mosaic.ini** file.

The **mosaic.ini** file is organized into sections, each of which performs one specific function, such as describing the fonts used. Section names are enclosed in brackets, for example: [Normal Font]. You can use any ASCII editor to edit the **mosaic.ini** file.

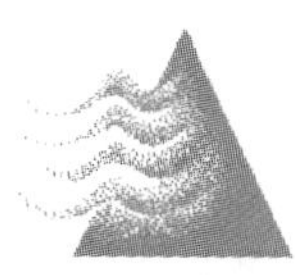

TIP
Make a backup copy of **mosaic.ini** before trying to edit it.

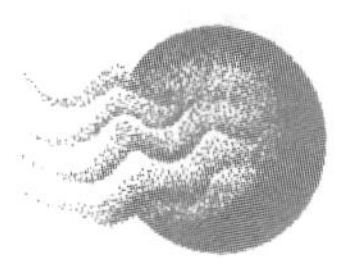

NOTE
Make sure that the directory, **c:\windows** by default, refers to the version of the **mosaic.ini** file associated with the latest version of Mosaic.

Syntax

The **mosaic.ini** file follows a set of simple rules. Its sections appear in a predetermined order. Within a given section, entries are coded on a single line.

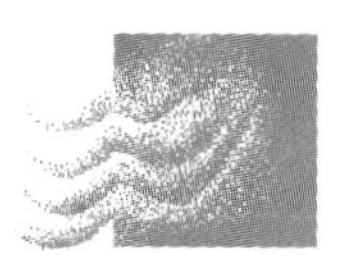

CAUTION
In this text, long entries may appear on two lines because of typesetting considerations.

Entries are of the form:

```
left side=right side
```

where *left side* is a variable such as *Autoload Home Page,* and *right side* is a value such as *yes* or a numeric value. A sample statement follows:

```
Autoload Home Page=yes
```

A statement beginning with the key word rem is a *comment,* an explanation or other nonexecutable statement. Comments do not directly affect Mosaic, but make it easier for people to understand and modify the **mosaic.ini** file. Consider the following statements taken from the Servers section.

```
SMTP_Server="ftp.ncsa.uiuc.edu"
rem=We know the above server will usually exist...
```

The comment supplies information about this server.

Main Section

The Main section is the first section of the **mosaic.ini** file. It contains basic information such as the user's e-mail address. This section describes selected entries in the Main section and explains how to change them. Here are the complete contents of the Main section:

```
[Main]
E-mail="put_your_email@here"
Autoload Home Page=yes
Home Page=http://www.ncsa.uiuc.edu/SDG/Software
          /WinMosaic/HomePage.html
Help Page=http://www.ncsa.uiuc.edu/SDG/Software
          /WinMosaic/Docs/WMosTOC.html
FAQ Page=http://www.ncsa.uiuc.edu/SDG/Software
          /WinMosaic/FAQ.html
Bug list=http://www.ncsa.uiuc.edu/SDG/Software
          /WinMosaic/Bugs.html
Feature Page=http://www.ncsa.uiuc.edu/SDG/Software
          /WinMosaic/Features.html
Display Inline Images=yes
```

```
Dump memory blocks=no
Grey Background=yes
Fancy Rules=yes
Round List Bullets=yes
Current Hotlist=Starting Points
Anchor Underline=yes
Anchor Cursor=yes
Show URLs=yes
Extended FTP=yes
Toolbar=yes
Status bar=yes
Title/URL bar=yes
Use 8-bit Sound=no
```

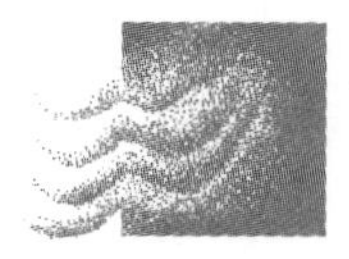

CAUTION

You cannot change section names.

The e-mail Variable

Chapter 5 showed how to specify your e-mail address in the **mosaic.ini** file. If you haven't yet done so, set your e-mail address now with an entry such as

```
E-mail="jradin@hookup.net"
```

The Autoload Home Page Variable

The *Autoload Home Page* variable specifies whether Mosaic automatically loads the Home Page (specified in the next variable) at system startup. The default value is no. Beginners may wish to see the Home Page without specifically requesting it. However, loading the Home Page takes time, especially when transmitted from a busy remote site such as the NCSA.

The Home Page Variable

The *Home Page* variable contains the address of your Home Page, the Mosaic startup document. The default value is

```
Home Page=http://www.ncsa.uiuc.edu/SDG/Software
         /WinMosaic/HomePage.html
```

Chapters 5 and 6 showed you how to create Home Pages. You can replace the NCSA Home Page with your own by using a command such as the following:

```
Home Page=file:///c¦/windows/user.htm
```

The Help Page Variable

The *Help Page* variable specifies the location of the Mosaic provider's Help Page. The file entry for NCSA Mosaic is

```
Help Page=http://www.ncsa.uiuc.edu/SDG/Software
          /WinMosaic/Docs/WMosTOC.html
```

By using the procedure described in the previous section, you can place an NCSA Help Page hyperlink in your own Home Page.

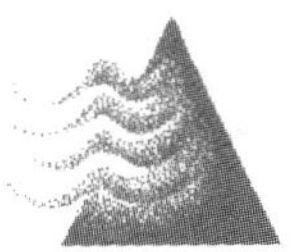

TIP

Users requiring help don't like waiting for the system to access a remote Help file. While it's a lot of work, it can be a good idea to download the remote Help Pages, copy it to a local file, and then place the local file hyperlink in your own Home Page. Because the Help Page refers to remote gif files, you must create a local copy of these files and then replace each reference to a remote gif file by a reference to the local file.

The FAQ Page Variable

The *FAQ Page* variable specifies the location of the Mosaic provider's FAQ (Frequently Asked Questions) Page. The file entry for NCSA Mosaic is

```
FAQ Page=http://www.ncsa.uiuc.edu/SDG/Software
          /WinMosaic/FAQ.html
```

Selecting FAQ Page from the Help menu generates a screen similar to the one shown in Figure 7-1.

The Bug List Variable

The *Bug list* variable specifies the location of the Mosaic provider's Bug List. The file entry for NCSA Mosaic is

```
Bug List=http://www.ncsa.uiuc.edu/SDG/Software
          /WinMosaic/Bugs.html
```

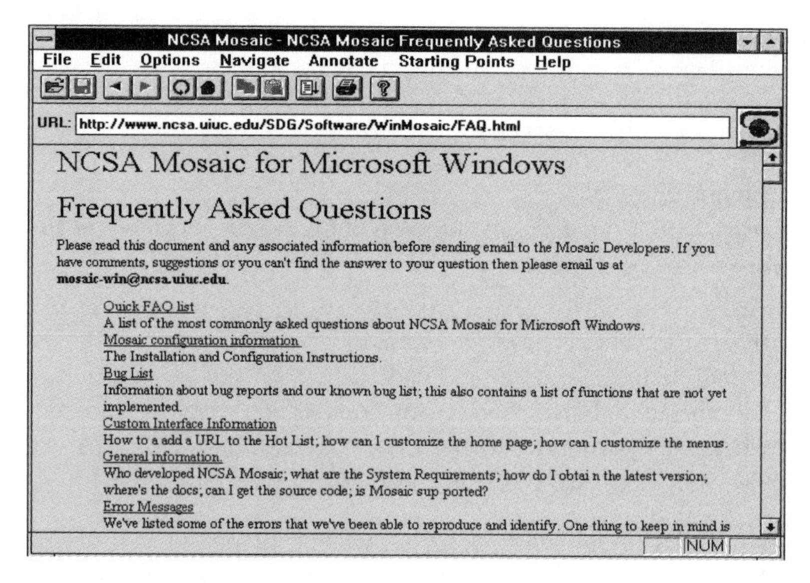

FIGURE 7-1. *FAQs (Frequently Asked Questions)*

Selecting Bug List from the Help menu generates a screen similar to the one shown in Figure 7-2.

The Feature Page Variable

The *Feature Page* variable specifies the location of the Mosaic provider's Feature Page, which contains general information about the new features and bugs fixed. The file entry for NCSA Mosaic is

```
Feature Page=http://www.ncsa.uiuc.edu/SDG/Software
        /WinMosaic/Features.html
```

Selecting Feature Page from the Help menu generates a screen similar to the one shown in Figure 7-3.

The Display Inline Images Variable

The *Display Inline Images* variable controls whether graphical images are transmitted with the document in which they appear. The default value is yes.

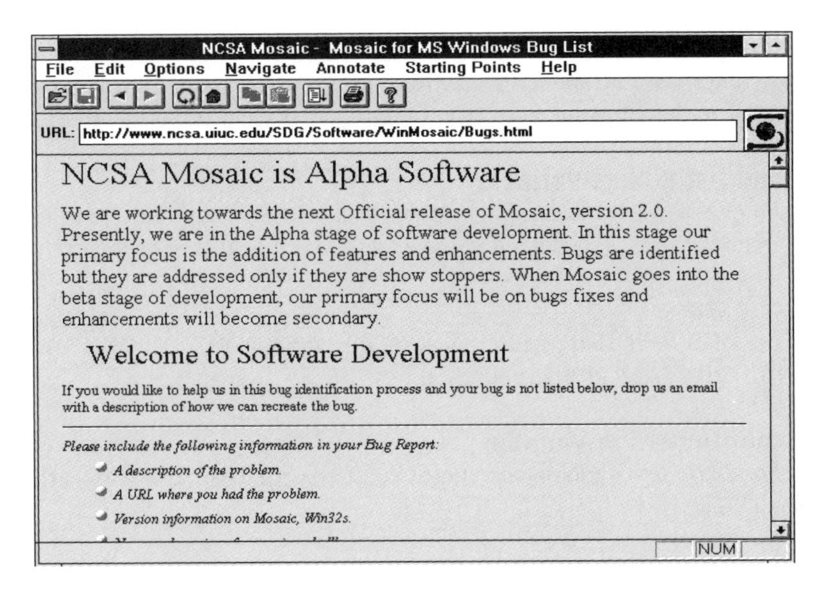

FIGURE 7-2. *A bug list*

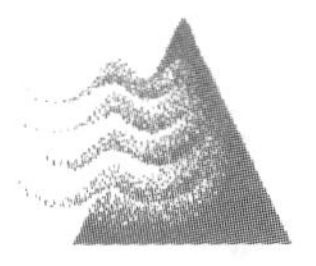

TIP

One way of improving Mosaic response time is to set this variable to no. In this case, the NCSA logo replaces each image as shown in Figure 4-14. To see the image, click on the logo with the right mouse button. To follow the associated hyperlink, click on the logo with the left mouse button.

The Grey Background Variable

The *Grey Background* variable specifies the background window color. The default value is yes, which results in a gray background. A no value results in a white background.

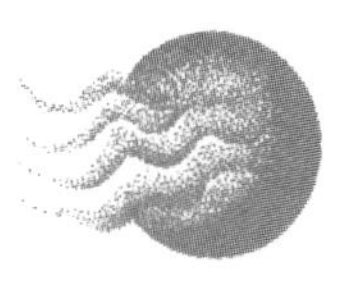

NOTE

All figures in this text were created with a gray background. As the old saying goes, "If it ain't broke, don't fix it."

The Fancy Rules Variable

The *Fancy Rules* variable specifies how to draw horizontal rules (lines) in the documents. The default value is yes for three-dimensional horizontal lines.

The Round List Bullets Variable

The *Round List Bullets* variable specifies whether to draw round or line bullets (dashes) in documents. The default value is yes, which results in round bullets.

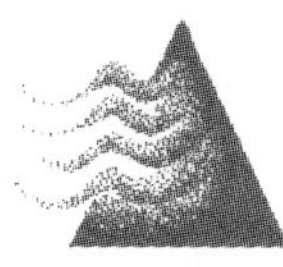

TIP
For increased performance, set this variable to no. It takes less time to draw dashes than round bullets.

The Anchor Underline Variable

The *Anchor Underline* variable specifies whether or not to underline hyperlinks. The default value is yes.

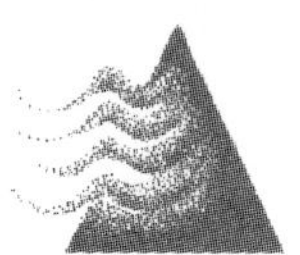

TIP
If you have a color monitor you can set the Anchor Underline value to no to increase system performance.

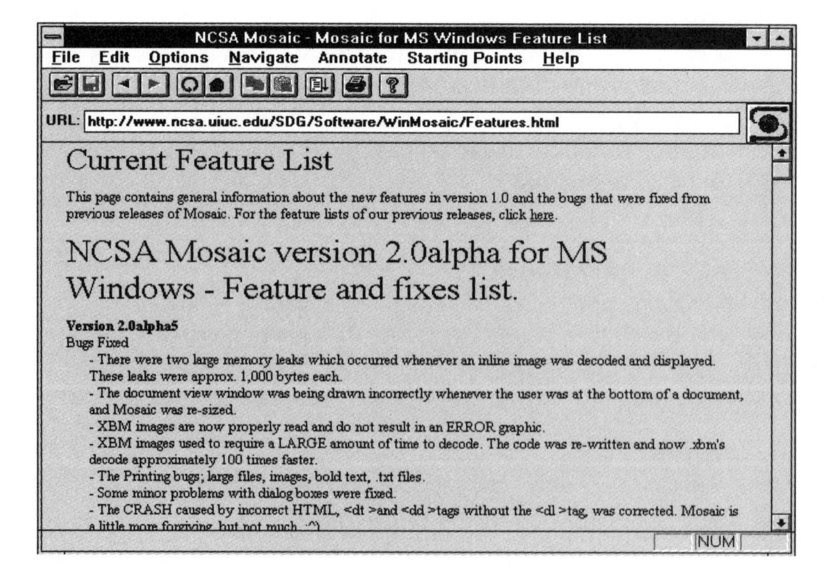

FIGURE 7-3. *A Mosaic feature list*

The Anchor Cursor Variable

The *Anchor Cursor* variable controls whether or not the hyperlink color is specified by the *Anchor Color* variable in the Settings section (which is described later). The default value is yes, meaning that the hyperlink color is specified by the *Anchor Color* variable.

The Show URLs Variable

The *Show URLs* variable controls whether the URL (Uniform Resource Locator) is displayed in the status window, provided that the *Status bar* variable is also set to yes. The default values of the *Show URLs* and *Status bar* variables are yes.

The Extended FTP Variable

The *Extended FTP* variable controls whether the file list displays file sizes in parentheses. The default value is yes (meaning that file sizes are displayed), and it's good to keep it this way.

The Toolbar Variable

The *Toolbar* variable controls whether the toolbar is displayed at the top of the Mosaic window. The default value is yes, which means that the toolbar is displayed. Figure 4-20 shows a screen generated when the *Toolbar* variable is set to no, via the menu.

The Status bar Variable

The *Status bar* variable controls whether the status bar is displayed at the bottom of the Mosaic window. The default value is yes, meaning that the status bar is displayed. The *Status bar* variable is used in conjunction with the *Show URLs* variable discussed earlier.

The Title/URL bar Variable

The *Title/URL bar* variable controls whether the Document Title, Document URL, and NCSA logo are displayed in the Mosaic window. The default value is yes, meaning that all of these elements are displayed. Figure 4-21 shows a screen when the *Title/URL bar* variable is set to no.

Settings Section

The Settings section specifies the color of hyperlinks. If you get tired of the standard blue, you can change it. To change to green for example, you could enter the settings **0, 255, 0**. The default section appears here:

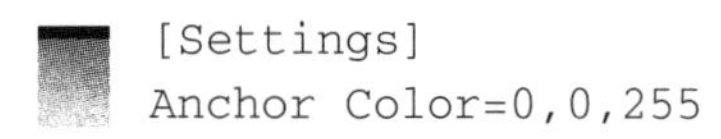

```
[Settings]
Anchor Color=0,0,255
```

The Anchor Color Variable

The *Anchor Color* variable specifies the *anchor color*, the color of hyperlink anchors in an NCSA Mosaic document. Set this color to any valid RGB (red, green, blue) combination. These three values each range from 0 to 255, and are separated by commas. As mentioned, the default color is blue (0,0,255).

Main Window Section

The Main Window section contains four variables that define the size and cursor location of the Mosaic application window. In general, there is no need to change these values, which are shown here:

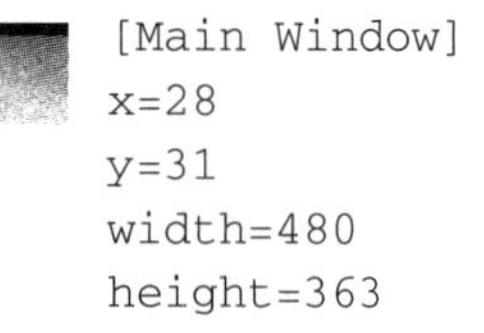

```
[Main Window]
x=28
y=31
width=480
height=363
```

Mail Section

The Mail section contains a single variable, the *Default Title* variable, which is described next.

```
[Mail]
Default Title="WinMosaic auto-mail feedback"
```

The Default Title Variable

The *Default Title* variable specifies the subject title for e-mail messages sent to Mosaic developers, perhaps via the Mail to Developers option on the Help menu. The default value is WinMosaic auto-mail feedback. There is no need to change this variable.

Services Section

The Services section defines server names for different types of communications. Here are its entries:

```
NNTP_Server="news.cso.uiuc.edu"
SMTP_Server="ftp.ncsa.uiuc.edu"
rem=We know the above server will usually exist.
    Change it if you have a local SMTP server.
```

The individual entries are discussed next.

The NNTP_Server Variable

The *NNTP_Server* (Network News Transport Protocol) variable specifies the server for Mosaic's news support. The NCSA Mosaic default server is **news.cso.uiuc.edu**, the University of Illinois' NNTP server. Change this value if your information provider uses a different server—for example:

```
NNTP_Server="nntp.ott.hookup.net"
```

The SMTP_Server

The *SMTP_Server* (Simple Mail Transport Protocol) variable designates the server used to send mail to Mosaic developers via the Help menu. The NCSA Mosaic default value is **ftp.ncsa.uiuc.edu**, the NCSA FTP server. Change this value if your information provider uses a different server—for example:

```
SMTP_Server="mail.ott.hookup.net"
```

Viewers Section

Appendix B discusses in detail how to modify the Viewers section in order to use external viewers and players. Viewers and players are programs that allow you to view or play external files. For example, you may use the program **lview31** to view images. Our modified Viewers section appears next. Yours will be somewhat different, as we explain later.

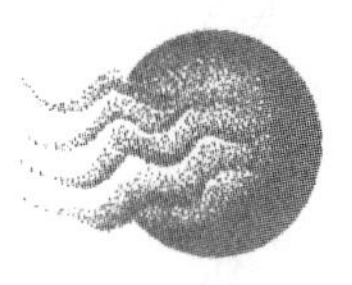

NOTE
You won't need to change most of these settings.

```
[Viewers]
TYPE0="audio/wav"
TYPE1="application/postscript"
TYPE2="image/gif"
TYPE3="image/jpeg"
TYPE4="video/mpeg"
TYPE5="video/quicktime"
TYPE6="video/msvideo"
TYPE7="application/x-rtf"
TYPE8="audio/x-midi"
TYPE9="application/zip"
rem TYPE9="audio/basic"
application/postscript="d:\gs\gswin -Id:\gs %ls"
rem image/gif="c:\windows\apps\lview31 %ls"
image/gif="d:\wingif\wingif %ls"
image/jpeg="c:\windows\apps\lview31 %ls"
video/mpeg="c:\windows\apps\mpegplay %ls"
rem video/quicktime="C:\WINAPPS\QTW\bin\player.exe %ls"
rem video/msvideo="mplayer %ls"
audio/wav="c:\windows\apps\mpegplay %ls"
audio/x-midi="c:\windows\apps\mpegplay %ls"
application/x-rtf="write %ls"
application/zip="C:\PKUNZIP.EXE %ls"
audio/basic="notepad %ls"
text/plain="notepad %ls"
text/html="notepad %ls"
telnet="d:\netmanag\telnet.exe"
```

The preceding *TYPE* statements are the ones from the initial **mosaic.ini** file. However, we modified many of the suggested directories. For example, the first directory entry in the initial **mosaic.ini** file was

```
application/postscript="ghostview % ls"
```

which we changed to

```
application/postscript="d:\gs\gswin -Id:\gs %ls"
```

In other words, the Postscript application gswin is found in the **d:\gs** directory instead of the standard Postscript application ghostview found in the default **c:\windows** directory. (These applications allow you to view and print Postscript files.) Change directories as required. Presently the **wingif** file located in the **d:\wingif** directory is used to view gif (image) files. To reactivate the initially installed

lview31 viewer found in the **c:\windows\apps** directory, simply remove the rem from the first image/gif= statement and place it before the second image/gif= statement.

```
image/gif="c:\windows\apps\lview31 %ls"
rem image/gif="d:\wingif\wingif %ls"
```

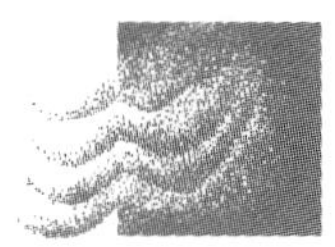

CAUTION
At any given time, Mosaic can only use a single viewer for any given file type such as gif files.

The final entry in the Viewers section specifies the complete path of the communications package used. The default value is

```
telnet="c:\trumpet\telw.exe
```

specifying the popular shareware communications software Trumpet.

Because we are using the popular commercial communications software Chameleon, we changed the setting to:

```
telnet="d:\netmanag\telnet.exe"
```

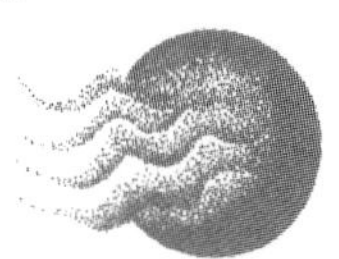

NOTE
Both Chameleon and Trumpet are discussed in detail in Appendix B.

Suffixes Section

Appendix B discusses in detail how to modify the Suffixes section in order to use external viewers and players. Our Suffixes section appears next. Yours will be probably be somewhat different.

```
[Suffixes]
application/postscript=.ps,.eps,.ai,.ps
application/zip=.zip
text/html=.html, htm
text/plain=.txt
application/x-rtf=.rtf,.wri
audio/wav=.wave,.wav,.WAV
audio/x-midi=.mid
image/x-tiff=.tiff,.tif
```

```
image/jpeg=.jpeg,.jpe,.jpg
video/mpeg=.mpeg,.mpe,.mpg
video/qtime=.mov
video/msvideo=.avi
```

In this Suffixes section, for example, files with extensions **.ps**, **.eps**, **.ai**, or **.ps** are Postscript applications.

Annotations Section

You use the Annotations section to attach personal or group comments to documents accessed via Mosaic. This section contains the following entries for NCSA Mosaic:

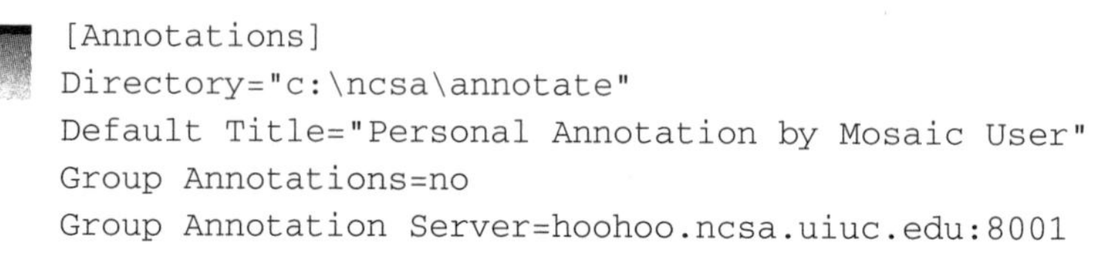

```
[Annotations]
Directory="c:\ncsa\annotate"
Default Title="Personal Annotation by Mosaic User"
Group Annotations=no
Group Annotation Server=hoohoo.ncsa.uiuc.edu:8001
```

Personal annotations are saved on your computer; group annotations are saved on the networked computer designated by the *Group Annotation Server* entry. Whenever you access a document, you will see your personal annotations. Whenever group members access the document, they will see the group annotations if these annotations are not disabled.

The Directory Variable

The *Directory* variable specifies the directory on your local hard disk where personal annotations are stored. The default value for NCSA Mosaic is "c:\ncsa\annotate". We changed this value to "d:\annotate".

The Default Title Variable

The *Default Title* variable indicates the title that will appear in the Annotate Window, as shown in Figure 7-4.

The Group Annotations Variable

The *Group Annotations* variable specifies whether an individual will see group annotations (yes) or personal annotations (no). Personal annotations are maintained on an individual's system and are available only to him or her. Group annotations are maintained on the group annotation server (see the next variable) and are available to all members of a specified group.

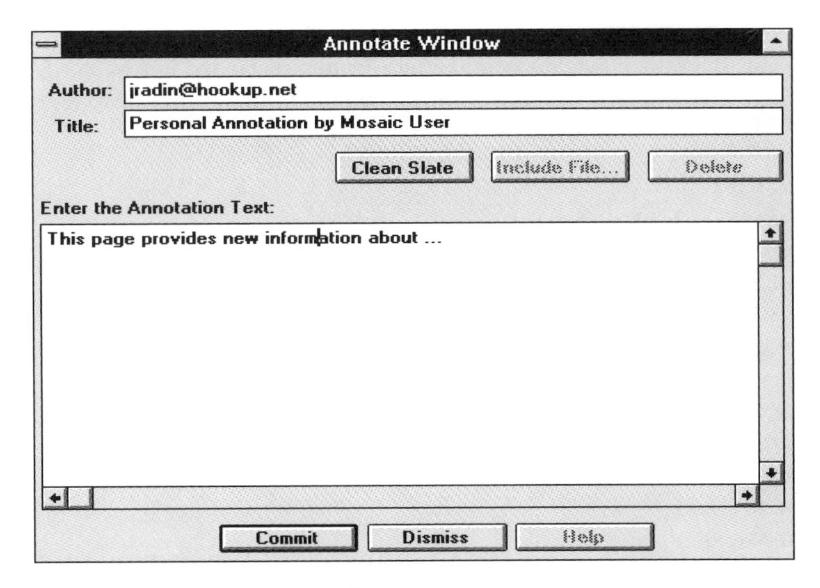

FIGURE 7-4. *The Annotate Window*

The Group Annotation Server

The *Group Annotation Server* entry points to your group annotation server, in this case, **hoohoo.ncsa.uiuc.edu:8001**.

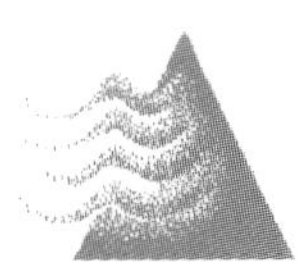

TIP
Specify a local server if you plan to make extensive use of group annotations.

User Menu Sections

The User Menu sections specify a maximum of 20 user-configured menus that provide easy access to interesting URLs. These menus are either top-level or pop-out menus. *Top-level menus* appear in the main menu bar. *Pop-out menus* appear when you select one of the top-level menus. Each menu can contain up to 40 entries, including menus defined within the given menu. Next we will examine one of our user menus in detail.

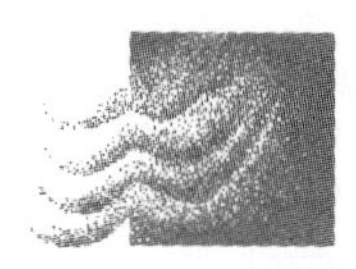

CAUTION
Use the Navigate, Menu Editor feature discussed in Chapter 8 to edit menus. Do not edit the **mosaic.ini** file directly; errors will be costly and hard to repair.

```
[User Menu1]
Menu_Name=Starting Points
Menu_Type=TOPLEVEL
Item1=Starting Points Document,http://www.ncsa.uiuc.edu/SDG
      /Software/Mosaic/StartingPoints/NetworkStartingPoints.html
Item2=NCSA Mosaic Demo Document,http://www.ncsa.uiuc.edu
      /demoweb/demo.html
Item3=NCSA Mosaic's 'What's New' Page,http://www.ncsa.uiuc.edu
      /SDG/Software/Mosaic/Docs/whats-new.html
Item4=MENU,User Menu2
Item5=MENU,User Menu3
Item6=MENU,User Menu4
Item7=Finger Gateway,http://cs.indiana.edu/finger/gateway
Item8=Whois Gateway,gopher://sipb.mit.edu:70
       /1B%3aInternet%20whois%20servers
Item9=MENU,User Menu5
```

The first line of a user-configured menu specifies the menu number, in this case

```
[User Menu1]
```

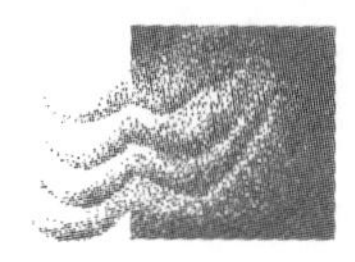

CAUTION
User menus must be specified in numeric order.

The next line specifies the menu name, in this case

```
Menu_Name=Starting Points
```

Then appears a line specifying the menu type, but only for top-level menus. User Menu1 contains the entry

```
Menu_Type=TOPLEVEL
```

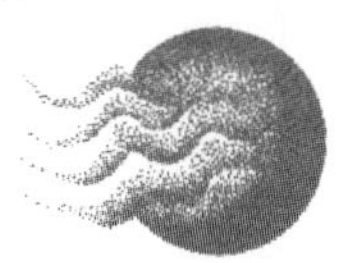

NOTE
A pop-out menu, one that does not appear in the menu bar, does not contain a *Menu_Type=* entry.

The next line is a bit more complicated.

```
Item1=Starting Points Document,http://www.ncsa.uiuc.edu/SDG
       /Software/Mosaic/StartingPoints/NetworkStartingPoints.html
```

The first menu item specifies the menu entry Starting Points Document (that appears on the screen), a comma, and the URL (its address) **http://www.ncsa.uiuc.edu /SDG/Software/Mosaic/StartingPoints/NetworkStartingPoints.html**.

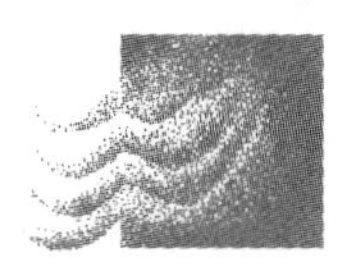

CAUTION
Within each menu, items must be specified in numerical order.

In the User Menu1 section section, *Item2* and *Item3* are similar to *Item1*. *Item4*, however, clearly has a different syntax.

```
Item4=MENU,User Menu2
```

This line, specifying a pop-out menu, contains the word *MENU*, a comma, and the number of a pop-out menu, defined elsewhere within the user menus section. Chapter 2 shows an example of the Starting Points menu. This completes the essentials of the User Menu syntax.

Hotlist Section

The Hotlist section stores *QUICKLIST URLs*, a list of URLs not displayed as a pop-out menu. This list is displayed when you choose Open URL from the File menu and choose QUICKLIST from the Current Hotlist list box. You can select any listed URL by scrolling to it, clicking on it, and clicking the OK button to confirm. Figure 7-5 shows our QUICKLIST. Let's look at the first few entries in the Hotlist section.

```
[Hotlist]
URL0=*Local C: Drive,file:///c¦/
URL1=*Mosaic Home Page,http://www.ncsa.uiuc.edu/SDG
       /Software/Mosaic/NCSAMosaicHome.html
URL2=*Mosaic Demo Page,http://www.ncsa.uiuc.edu
      /demoweb/demo.html
URL3=http://cs.indiana.edu/cstr/search
```

The first entry, URL0=, indicates the location of the Hotlist, in this case local drive C.
The entry URL1= indicates an optional Hotlist identifier, Mosaic Home Page, a comma, and the URL.

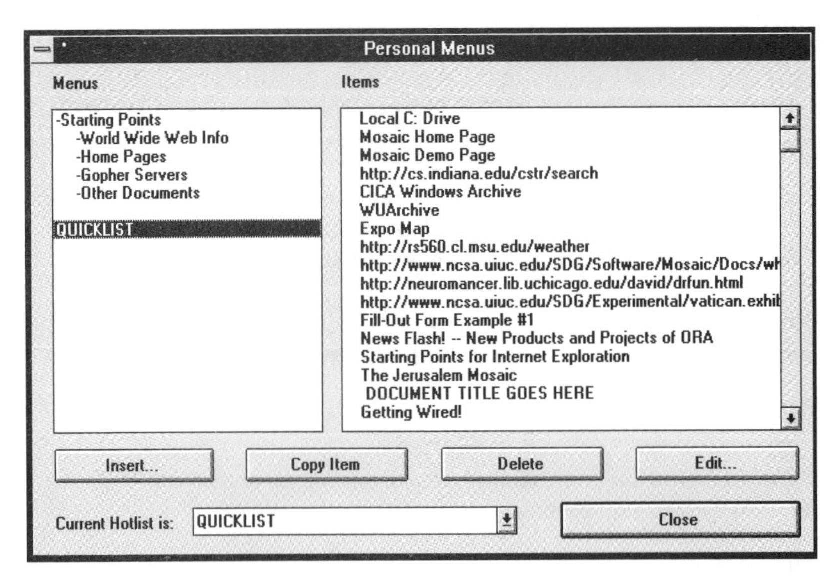

FIGURE 7-5. *Generating the QUICKLIST*

The entry URL3= indicates the URL but does not include a Hotlist identifier or a comma. This is an entry that did not appear on the original Hotlist, but was added with the procedure described next.

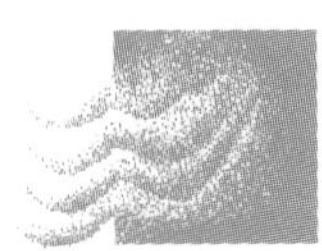

CAUTION
Use the Navigate, Menu Editor feature discussed in Chapter 8 to edit the QUICKLIST URLs. Do not edit the **mosaic.ini** file directly.

Adding a URL to the QUICKLIST

To add a URL to the QUICKLIST, pull down the Navigate menu, choose Menu Editor, and select QUICKLIST from the Current Hotlist list box. Finally, click on the Insert button to generate a dialog box similar to the one shown here:

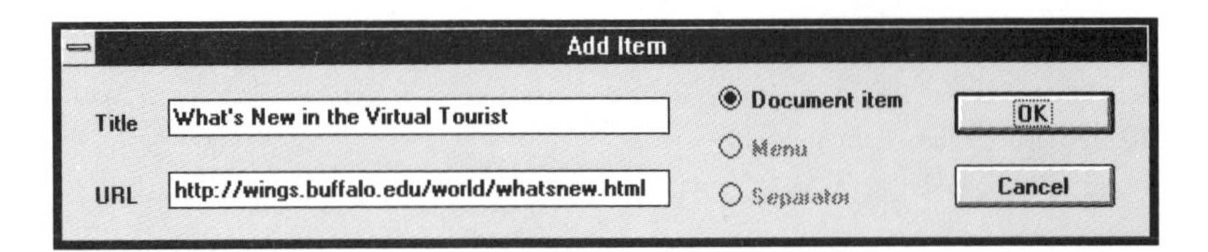

Enter either the document title or the URL (or verify the existing ones). Click on OK to confirm.

You can add URLs to the QUICKLIST or other Hotlists. To change menus, select the desired menu from the Current Hotlist list box in the Personal Menus dialog box. Use the Insert, Copy Item, Delete, and Edit buttons to control the contents of that menu.

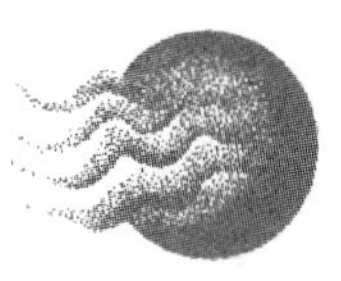

NOTE
The easiest way to add a file to your Current Hotlist is to select Add Current to Hotlist from the Navigate menu.

Document Caching Section

The Document Caching section specifies cache parameters that reduce the frequency of reaccessing the network to retrieve a recently viewed document from memory. The entries used on our 8MB computer are shown here:

```
[Document Caching]
Type=Number
Number=4
```

The Number Variable

The *Number* variable denotes the number of documents in the memory (cache). The default value is two documents.

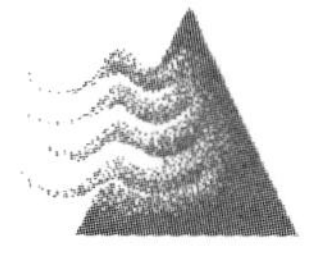

TIP
If your computer has lots of memory, you can increase the cache number. If your computer has 4MB of memory or less, you can decrease the cache number.

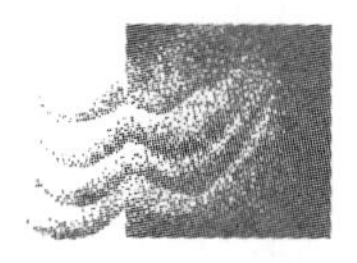

CAUTION
Do not modify the Type entry.

Font Sections

The **mosaic.ini** file contains several sections that define fonts. These include the Normal Font section, several heading font sections, the Menu Font section, the Dir Font section, the Address Font section, the BlockQuote Font section, the Example Font section, the Preformatted Font section, and the Listing Font section.

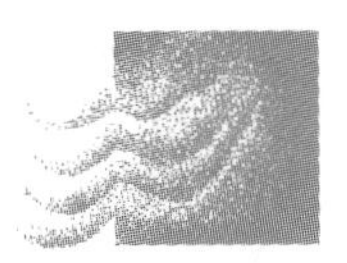

CAUTION
Do not edit the font section directly unless you want permanent changes. Make font changes from the Options menu.

For example, our Normal Font section includes the following entries:

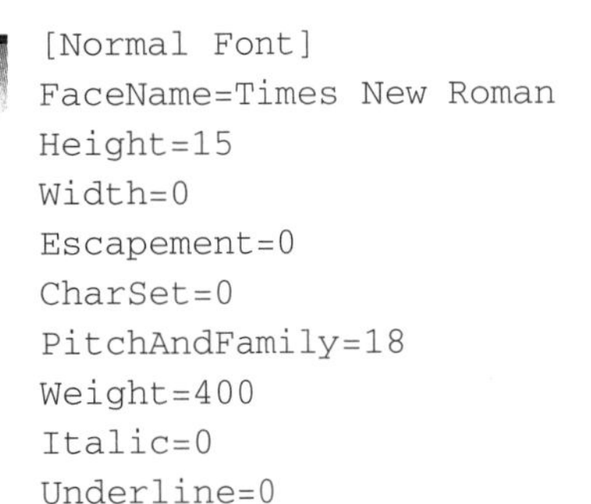

```
[Normal Font]
FaceName=Times New Roman
Height=15
Width=0
Escapement=0
CharSet=0
PitchAndFamily=18
Weight=400
Italic=0
Underline=0
```

The Examples Font section that defines the way examples appear on the screen contains three different entries:

```
FaceName=Courrier New
Height=13
PitchAndFamily=49
```

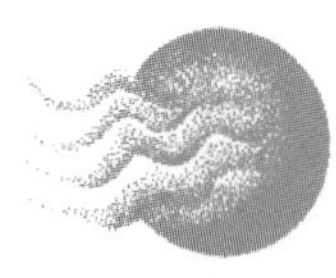

NOTE
Font design requires special skills. If you don't know what you're doing, stay with the default values. Chapter 4 shows how to use the menus to change the fonts.

Proxy Information Section

A proxy gateway allows Mosaic clients without Internet access to transfer network requests to a trusted agent who will access the Internet for them. At the time of this writing, proxy gateways are not commonly used.

Online Documentation

Mosaic provides online documentation in zip format. To use this documentation, follow these steps:

1. Move to the directory for file storage.

2. Download the **mosdoca5.zip** file, and unzip it. The unzipped file will include the file **wmostoc.htm**, which is the Home Page for Mosaic online documentation.
3. Copy this file to the chosen directory.
4. Access the file to generate a screen such as the one shown in Figure 7-6.
5. Scroll down and click, for example, on the Navigating the World Wide Web hyperlink to generate the documentation screen shown in Figure 7-7.
6. You can click on the arrow icons (the left arrow accesses the previous section) or text hyperlinks for more information.

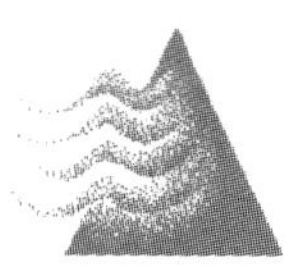

TIP

If you plan to make heavy use of Mosaic's online documentation, transfer all files, including the gif files (the arrow images) into a local directory. This will involve a lot of work, but users who need help don't want to wait for lengthy file transfers.

Configuring Mosaic for X Window (Unix)

Most users of the Unix version of Mosaic will make few, if any, changes to the configuration file **Mosaic**. In general, understanding this file requires familiarity

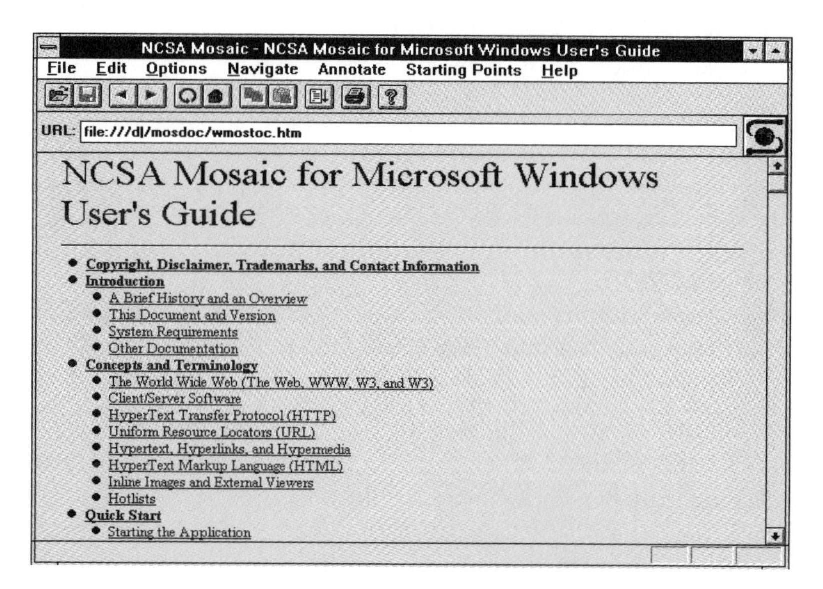

FIGURE 7-6. *Accessing the online documentation table of contents*

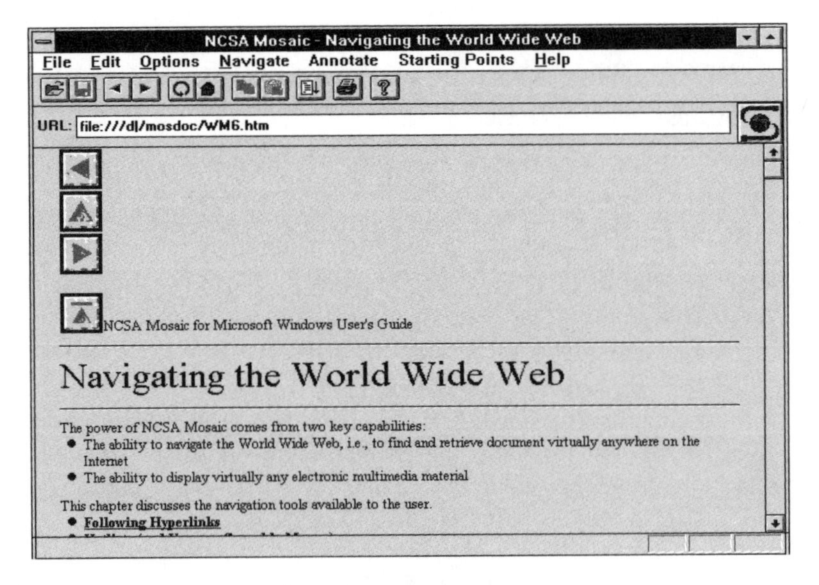

FIGURE 7-7. *Examining online documentation*

with the X Window system, the standard windowing system for Unix. If you want more details on this system, consult *X Window Inside and Out* by Levi Reiss and Joseph Radin (Osborne McGraw-Hill, 1992).

For example, consider the first line of the **Mosaic** file:

```
Mosaic*XmLabel*fontList:    -*-helvetica-bold-r-normal
        -*-14-*-*-*-*-*-iso8859-1
```

This line specifies the font used for labels in the windows. Don't change it.

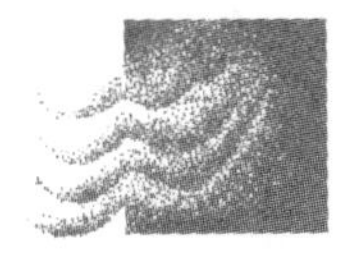

CAUTION

For typesetting reasons, some statements such as the above one appear on two lines. However, in the actual **Mosaic** file all entries must appear on a single line.

Some entries in the **Mosaic** file are similar in function to the associated variables in the **mosaic.ini** file. The value 1 or True corresponds to the value yes used in the **mosaic.ini** file. For example, the line

```
Mosaic*AnchorUnderlines:                    1
```

is similar to the line

```
Anchor Underline=yes
```

in the **mosaic.ini** file.

Of course, it is very easy to change this entry with an ASCII editor.

By now, you should be familiar with the reasons for setting up your own Home Page. To do so, modify the Mosaic*homeDocument: line in the **Mosaic** file to an entry such as the following:

```
Mosaic*homeDocument:    file://localhost/usr/users/joe/home.html
```

This sets up a Personal Home Page whose file name is **home.html** in the **/usr/users/joe** directory on the local computer defined as **localhost**.

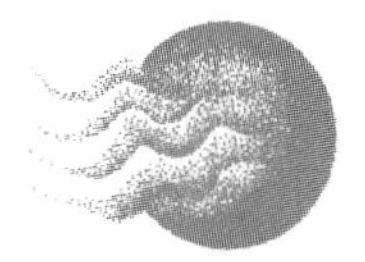

NOTE

The Unix version uses the extension .html in contrast to the Windows version extension .htm.

Mosaic Information Files

The **Mosaic** file is used to configure the Unix version of Mosaic. The following files contain valuable information about your Mosaic continuous status.

The .mosaic-global-history File

The **.mosaic-global-history** file is a list of all URLs activated since the last purge. To reactivate a document, simply click on it. An entry appears below:

```
http://english-server.hss.cmu.edu/Govt.html
       Wed Aug  3 16:52:02 1994
```

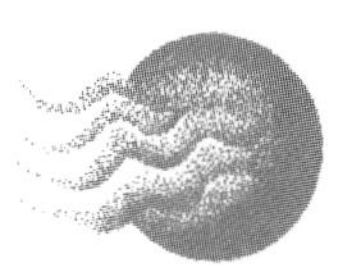

NOTE

For typesetting reasons, some entries appear on two lines. In the actual file all entries appear on a single line.

The .mosaic-hotlist-default File

The **.mosaic-hotlist-default** file contains Hotlist entries composed of a URL and a document title, as shown here:

```
http://info.cern.ch/hypertext/WWW/Daemon/User/Config
     /Overview.html Wed Jun 29 16:22:46 1994
Syntax of CERN server configuration (rule) file
```

The .mosaipid File

The **.mosaipid** file contains the process ID for an executing copy of Mosaic. The following procedure shows you how to kill Mosaic by using this file entry.

```
cat .mosaipid
```

to obtain the process ID.

```
kill -9 process_ID
```

The .mosaic-personal-annotations Directory

The **.mosaic-personal-annotations** directory contains personal annotations. To obtain these annotations, enter the following:

```
LOG
mosaic-personal-annotations
Personal
```

Troubleshooting

The following errors can occur on both the MS-Windows and Unix versions of Mosaic.

Failed DNS Lookup

The Failed DNS Lookup error could be caused by a few things. First, the IP number you supplied to your network software could be wrong. (The Domain Name Service (DNS) translates network names to IP addresses.) Second, the name you are trying to resolve may not be associated with an IP number—that is, it might be a typo or the name may not exist. Third, the DNS may be down.

No Menus

The No Menus error occurs when the **mosaic.ini** file is not in your Windows directory or when the environment variable defined in your **autoexec.bat** file is wrong. A correct value could be

```
set MOSAIC.INI=c:\complete_directory_path\mosaic.ini
```

HT Access Error

Clicking on a hyperlink before document transfer is completed generates an error message such as the following:

```
"http:// (rest of URL for document being transferred)":
      "Transfer Cancelled
```

If this occurs, click on the OK button and try again later.

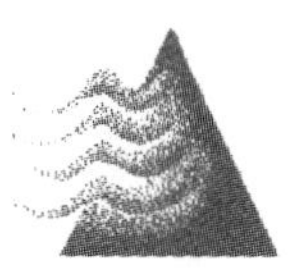

TIP
Wait until clicking on the scroll box at the right side of the screen changes the display before clicking on a hyperlink.

Mosaic Menus and Buttons

This concluding chapter takes you on a tour of the menus and toolbar for the MS-Windows version of Mosaic, and a tour of the similar menus and push buttons for X Window System (Unix) Mosaic.

Mosaic Menus

Chapters 2 and 3 explored the menus featured in the MS-Windows and X Window System versions of Mosaic. This chapter reviews some of the menu items that you learned about earlier and introduces the remaining selections. First you learn about the Windows menus; then you explore the X Window menus. While the two series of menus are discussed separately, they are quite similar. At the end of the chapter, you learn about the buttons available in Mosaic. Again, the Windows buttons are discussed first, followed by the X Window buttons.

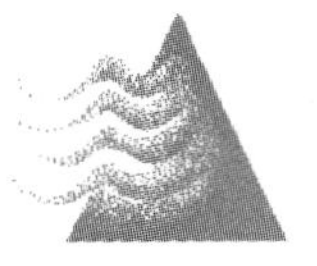

TIP
In general, use the menus for temporary specifications and the **mosaic.ini** file for more permanent specifications. For instance, if you want to change a font once, or only sporadically, you can use the menus. But if you want to change the font "for all time," you can modify the **mosaic.ini** file. (See Chapter 7 for additional details on the **mosaic.ini** file.)

Mosaic Main Menu for Windows

The main menu for the MS-Windows version of Mosaic contains seven top-level menus, as shown in Chapter 2. Selecting a top-level menu generates a pull-down menu of additional choices. These top-level menus include the File menu, the Edit menu, the Options menu, the Navigate menu, the Annotate menu, the Starting Points menu, and the Help menu.

The File Menu

You use the File menu to load documents directly over the Internet or from local disk—for example, when storing search results. The File menu includes ten options, which are described next.

- **Open URL...** provides access to a document whose URL (Uniform Resource Locator) you specify. You have done this often—for example, when you generated the list of commercial World Wide Web servers in the beginning of Chapter 4. This menu item also provides access to Hotlists via the Current Hotlist field, as shown in Chapter 4.
- **Open Local File...** opens a dialog box for specifying the local file name. The dialog box is a standard MS-Windows File Manager interface.
- **Save** saves the current document.
- **Save As...** saves the current document to a local file whose file name you specify. The available file formats are ASCII, HTML (hypertext format), and binary.

- **Save Preferences** saves the location and dimensions of the display window and any specifications you made via the Options menu.
- **Print...** displays a standard MS-Windows Print dialog box for printing the specified document.

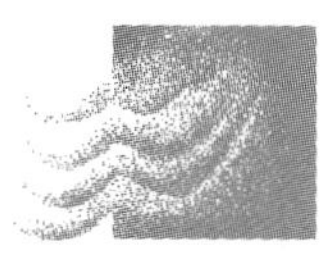

CAUTION
Scalable fonts are required for fully formatted printouts.

- **Print Preview** shows you what your printed document will look like, as shown in Figure 8-1. Judicious use of the Print Preview option can save you time and paper.
- **Print Setup...** generates a standard MS-Windows dialog box for specifying which printer to use, which paper size to use, and other printing options.
- **Document Source...** provides an internal view of your Mosaic document in either ASCII (plain text) or HTML (hypertext) format. Clicking on this menu option generates the two menus File and Edit. Click on File¦Save to save the current file to a local disk. Click on File¦Exit to close the document source window and return to the (usual) document view window. Click on Edit¦Copy to copy the selected text to the Clipboard.
- **Exit** lets you leave Mosaic.

The Edit Menu

You use the Edit menu to copy or find a document segment such as a *character string*, a series of characters. The Edit menu offers two selections, which are described next.

- **Copy** copies selected text to the Clipboard. You can choose from among Document Title, Document URL, source window, and text entry fields such as Open URL window. (This option is not fully implemented.)

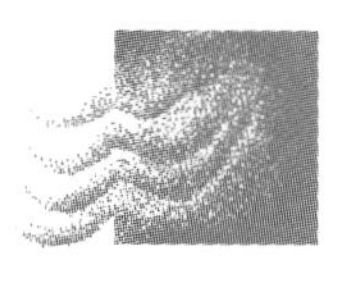

CAUTION
You cannot select text from the document display area.

- **Find...** displays a Find dialog box in which you enter a character string to search for. You should indicate whether or not you want the search to be case-sensitive.

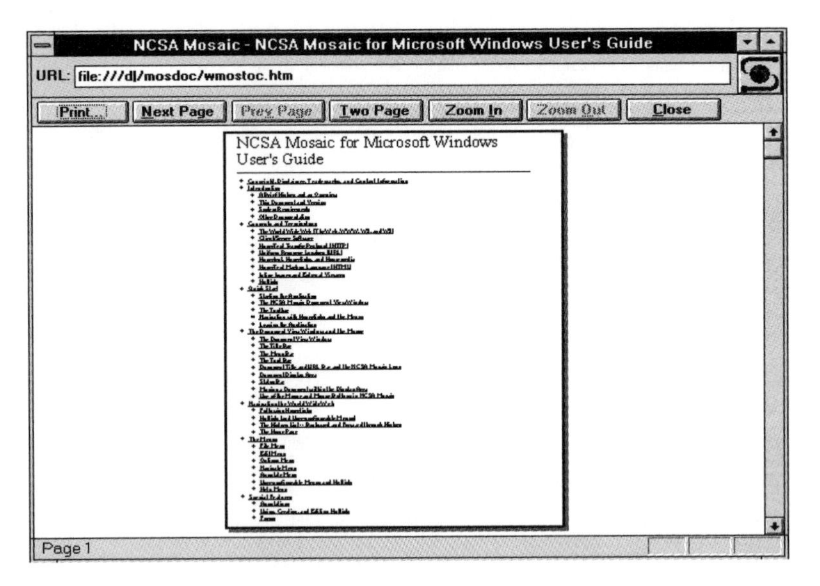

FIGURE 8-1. *Previewing a document*

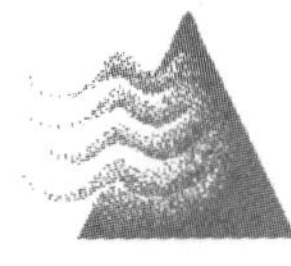

TIP

When working with long documents, use the Find feature instead of doing a manual search. Remember, your eyes can play tricks on you, especially when you are tired.

The Options Menu

You use the Options menu to customize the appearance of the screen. Many menu items are *toggles*: selecting them once turns them on (places a checkmark to their left) and selecting them again turns them off (removes the checkmark). For example, a checkmark to the left of Show Status Bar means the status bar is visible; no checkmark means it is hidden.

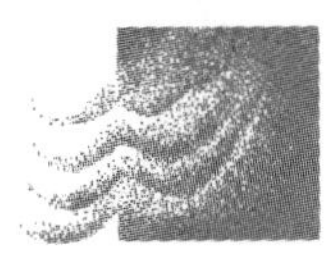

CAUTION

Check the state of a toggle option before clicking on it. If you click on a toggle item blindly, you may be unpleasantly surprised at the results.

The Options menu provides 12 options, as described here.

- **Load to Disk**, when activated, causes clicking on a hyperlink to display a dialog box for saving a document to a local disk instead of displaying it. Either specify the directory and file name or accept the default values. In Chapter 4, you used this option to save special images and movies.

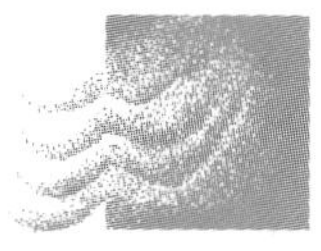

CAUTION
Disable the Load to Disk option immediately after loading the desired images to disk. Otherwise, all incoming files will be stored to disk. You won't see documents on the screen and you may run out of disk space. Long file names are truncated to meet MS-Windows file naming conventions. Only the first eight characters prior to the optional period are retained, only the first three characters after the period are retained, and all characters between multiple periods are discarded. For clarity, you should change file names that have been truncated.

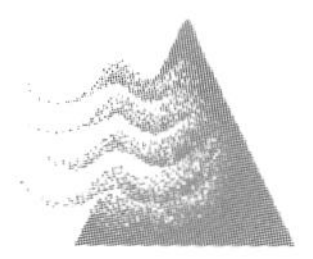

TIP
To save a single document, hold down the SHIFT key and click on the hyperlink.

- **Show Toolbar** shows or suppresses the toolbar, a handy bar near the top of the screen that includes icons that are mouse shortcuts for frequently executed commands. For instance, you can click on the file folder icon to open a file. You should keep the toolbar on, it provides a quick way of performing a number of useful tasks. Figure 4-20 shows a display in which the toolbar was suppressed.
- **Show Status Bar** shows or suppresses the status bar, a line at the bottom of the screen that displays information such as the number of bytes transferred and the Document URL. The status bar also indicates when the CAPS LOCK, NUM LOCK, or SCROLL LOCK key is on. It's a good idea to keep the status bar on, since it shows you the progress of document transmission and displays error messages.
- **Show Current URL/Title** shows or suppresses the document title and its Uniform Resource Locator. You should make sure that the Current URL/Title option is checked. This information is quite useful, especially if you get lost. Figure 4-21 shows a display in which the Current URL/Title was suppressed.

- **Show Anchor URLs** shows or suppresses the underlining or special color for hyperlinks such as World Wide Web. It's a wise policy to keep this toggle switch on. Hyperlinks are very important and should be readily visible.
- **Change Cursor Over Anchors** shows or suppresses a different cursor shape such as a hand when the cursor is positioned over a hyperlink. You should keep this toggle switch on since it's a good policy to do everything you can to make hyperlinks readily visible.
- **Extended FTP Directory Parsing** displays the file names and directory for a server when disabled; also displays file sizes in parentheses when enabled.
- **Display Inline Images** shows or suppresses images in the transmitted document. Suppressing the image reduces the transmission time. The image is replaced by an icon. Click on this icon with the right mouse button to load the image; click on it with the left mouse button to follow the associated hyperlink.
- **Show Group Annotations** shows or suppresses group annotations (comments made by other members of your workgroup). See the later section on the Annotate menu for additional details.
- **Use 8-bit sound** directs sound files to the 8-bit sound system hardware. Given Mosaic's commitment to keeping up with technology, future versions should offer more sophisticated audio capabilities.
- **Choose Font...** specifies the font type for one or more selected paragraphs, designated as Normal, Header (1 to 7), Menu Directory, Address, Block Quote, Example, Preformatted, and Listing. These designations refer to HTML elements discussed in Chapter 6. Figure 8-2 shows a sample font specification for the Header (originally defined as Times New Roman, Regular, 24). Figure 8-3 shows the Home Page with the new specifications. As you can see, the name to the right of the image appears in both bold and italics, as specified in Figure 8-2. You can also modify the **mosaic.ini** file if you want to redefine the fonts more permanently, as discussed in Chapter 7.
- **Debugging info** provides information that may help you determine what went wrong. Menu items include Info, Warnings, Critical Warnings, Errors, Trace, None, All, Old, and Dump object lists. Most users will not make use of this menu, except as directed by a specialist.

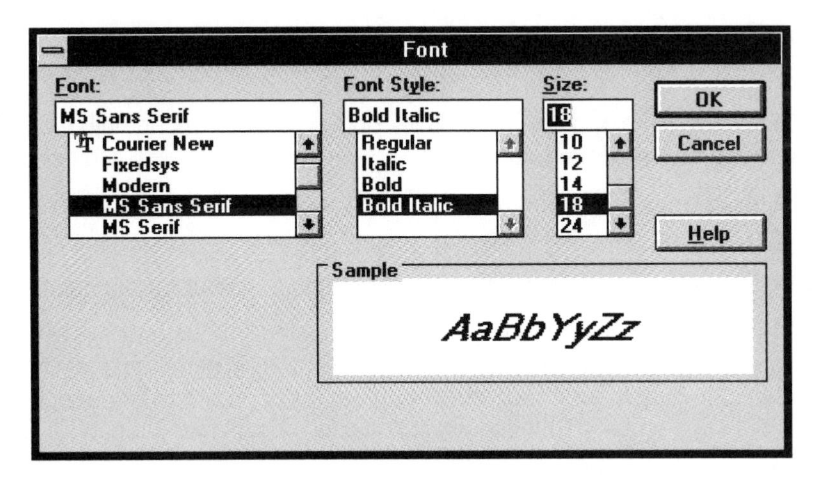

FIGURE 8-2. *Specifying the MS Sans Serif font*

The Navigate Menu

The Navigate menu lets you make your way efficiently across the Internet. As you learned in Chapters 4 and 5, you can access this menu repeatedly to get the

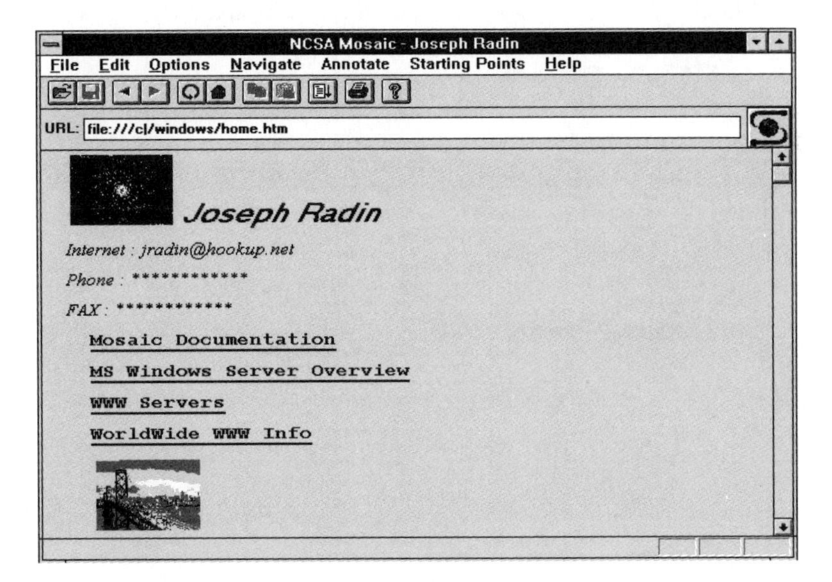

FIGURE 8-3. *Showing the new font*

information you want. The Navigate menu includes the seven options described here:

- **Back** retrieves the most recently accessed document. Use this repeatedly to access documents in reverse order.
- **Forward** is the opposite of the Back option. Use this menu item when you overshoot the desired document.
- **Reload** reloads the current document. Try using this menu item when you get an error loading a document.
- **Home** returns you to the Home Page specified in your **mosaic.ini** file.
- **History** displays a list of the URLs for documents in the order in which they were accessed during the current information gathering session. To select a entry in the history file, double-click on it, or click on it once and then click on Load in the toolbar. You used this feature often in Chapter 5.
- **Add Current to Hotlist...** places an interesting document in the Hotlist for rapid retrieval.
- **Menu Editor...** opens the Personal Menus window for editing user-configurable menus and the QUICKLIST, as shown in Figure 8-4. Use the Current Hotlist drop-down list box to select a different Hotlist. Click on Insert to display the Add Item dialog box shown in Figure 8-5. Fill in this dialog box and click on OK to add the URL to the QUICKLIST. Refer back to Chapter 7 for more information on the Hotlist and the QUICKLIST.

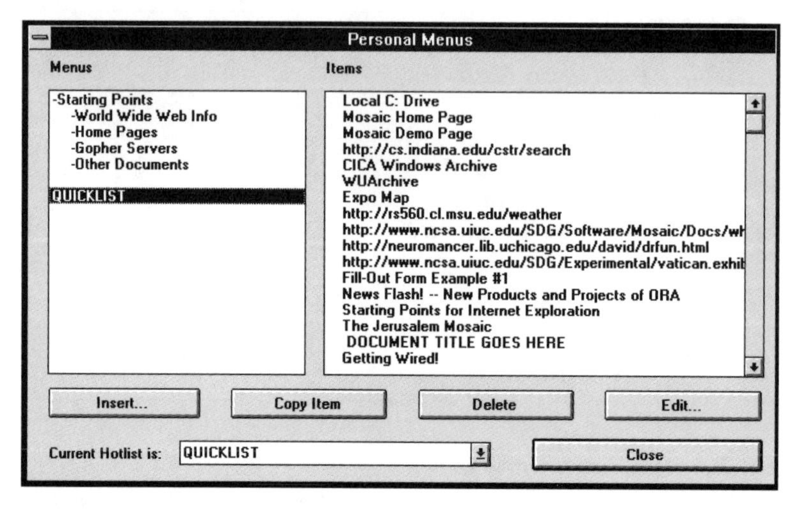

FIGURE 8-4. *Accessing the QUICKLIST*

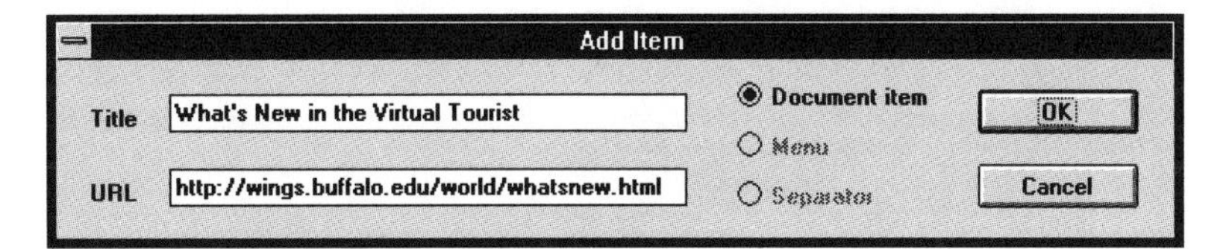

FIGURE 8-5. *Adding an item to the QUICKLIST*

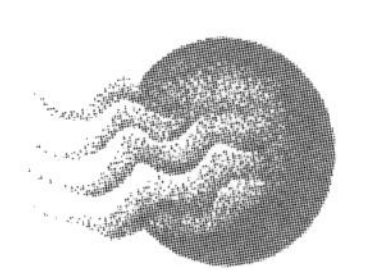

NOTE
In general, it's most efficient to first find an interesting document and then add its URL to the desired Hotlist without ever jotting it down on paper.

The Annotate Menu

The Annotate menu enables you to add personal or group comments to Mosaic documents for future reference, as specified in the Annotate section of the **mosaic.ini** file. The Annotate menu offers these three selections:

- **Annotate** displays the Annotate Window shown in Figure 8-6, in which you enter annotations. Figure 8-7 shows the hyperlink at the end of the annotated document. Click on this hyperlink to display the annotation as shown in Figure 8-8. The annotation includes the annotator's e-mail address in case further communication is necessary.
- **Edit this Annotation** enables you to modify the current annotation.
- **Delete this Annotation** lets you remove the current annotation.

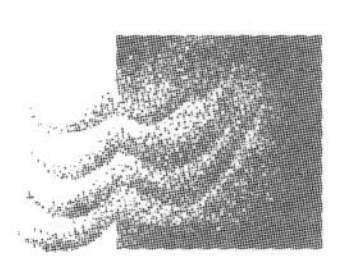

CAUTION
Delete annotations with care. There is no Undelete feature.

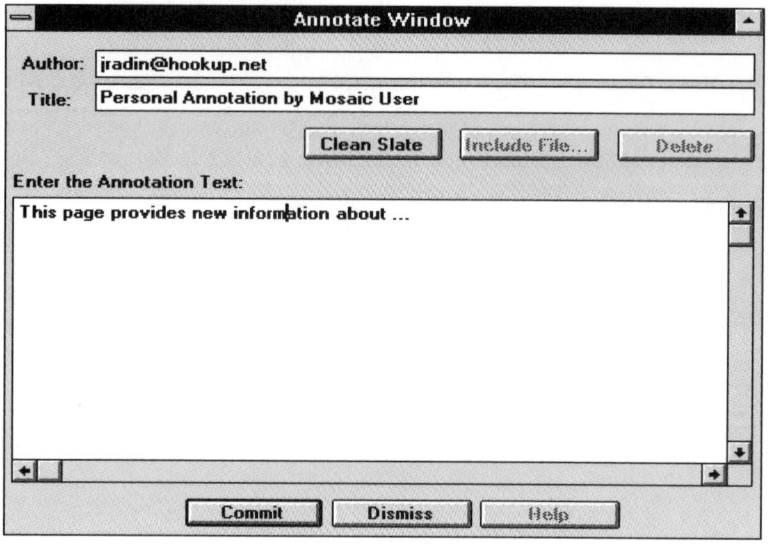

FIGURE 8-6. *The Annotate Window*

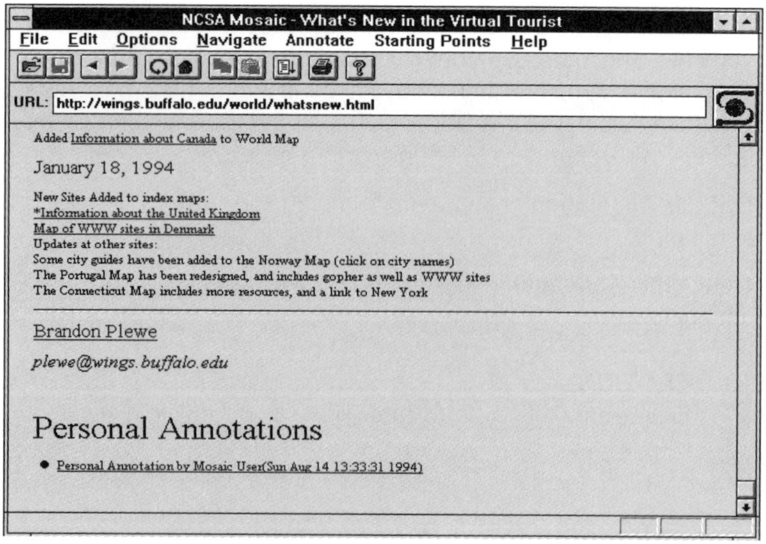

FIGURE 8-7. *A hyperlink to the annotation*

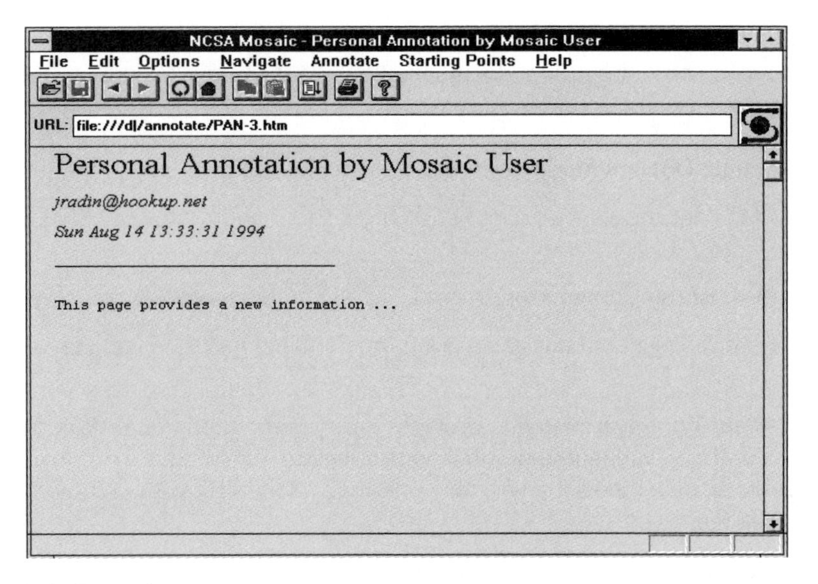

FIGURE 8-8. *The annotation*

The Starting Points Menu

No matter how familiar you are with the Internet, the Starting Points menu is a good place to begin searching for information. This menu provides at least the following useful hyperlinks, plus others along the way.

- **Starting Points Document** lists points for starting an unstructured search as in Chapter 5. It contains hyperlinks to many common Internet-based information resources.
- **NCSA Mosaic Demo Document** is an interactive hypermedia tour of Mosaic's capabilities.
- **NCSA Mosaic's What's New Page** covers recent changes and additions to the information available to Mosaic and the World Wide Web.

Each of these Starting Points leads to numerous rich information sources. If you haven't tried them yet, try them now.

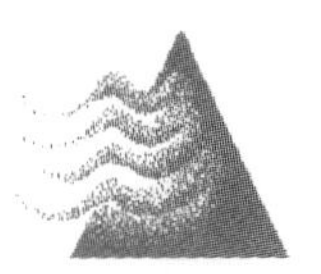

TIP

Mosaic can handle about 20 user-configurable menus and submenus. Use the Menu Editor discussed earlier (it's in the Navigate menu) to create and modify user-configurable menus so that you can easily get to commonly accessed URLs.

The Help Menu

The Help menu provides online documentation about Mosaic. It contains the following options:

- **Online Documentation** provides explanations at the click of a mouse.
- **FAQ Page** answers frequently asked questions about Mosaic, as shown in Figure 7-1.
- **Bug List** lists known Mosaic errors, as shown in Figure 7-2.
- **Feature Page** contains general information about Mosaic features, as shown in Figure 7-3.
- **About Windows Mosaic...** provides basic information about Windows Mosaic, including the product name, version number, a short copyright notice, developers' names, and an e-mail address for your comments and bug reports.
- **Mail to Developers...** opens a window in which you can compose and send messages to technical specialists.

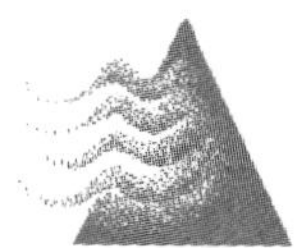

TIP

Before communicating with Mosaic developers, check the Online Documentation's FAQ Page, Bug List, and Features List discussed a moment ago. Doing so can save everybody, including yourself, valuable time and energy.

Mosaic Main Menu for X Window (Unix)

The main menu for the X Window version of Mosaic contains five top-level menus. Selecting a top-level menu displays a pull-down menu of additional choices. These top-level menus include the File menu, the Options menu, the Navigate menu, the Annotate menu, and the Help menu.

The File Menu

You use the File menu to load documents directly from the Internet or from local files—for example, when temporarily storing search results. The File menu includes 14 options, which are described next.

- **New Window** opens a new window.
- **Clone Window** opens a copy of an existing window.
- **Open URL...** opens a dialog box in which you enter the URL of the desired document.
- **Open Local...** opens a local document, one that can be accessed directly.
- **Reload Current** gets a copy of the current document if the Home Page has been changed.
- **Reload Images** gets a copy of graphics files linked to the current document if the Home Page has been changed.
- **Refresh Current** redraws the screen for the current document if the Home Page has been changed.
- **Find...** locates the next occurrence of the designated character string in the document. You can search forward, and can choose whether to make searches case sensitive.
- **View Source...** provides an internal view of your Mosaic document, but there are no editing features here.
- **Save As...** saves the current document with a file name that you specify and confirm. You can save the file in one of four formats: Plain Text, Formatted Text, Postscript, or HTML (Mosaic hypertext format).
- **Print...** prints a specified document in one of four formats: Plain Text, Formatted Text, Postscript, or HTML (hypertext format).

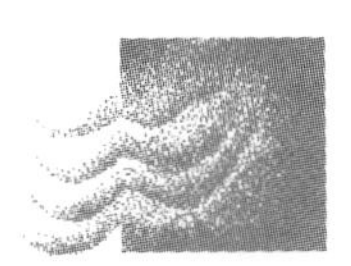

CAUTION
No print preview, so be careful.

- **Mail to...** lets you send mail to the developers of Mosaic.
- **Close Window** closes a window.

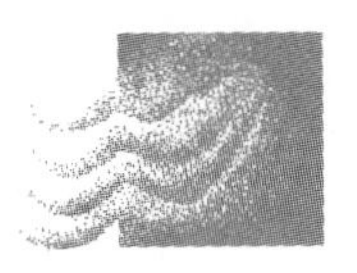

CAUTION
Closing the last window closes Mosaic without giving you a chance to change your mind.

- **Exit Program...** lets you leave Mosaic after confirmation.

The Options Menu

You use the Options menu to customize the appearance of the screen. Many options on this menu are toggles that you can turn either off (pushed out) or on (indented and perhaps in color) by selecting them. The Options menu includes the nine selections outlined here.

- **Fancy Selections**, when activated, preserves as much formatting (e.g. headers, tools) as possible with selections made from the X Window system cut and paste mechanism.
- **Load to Disk**, when activated, causes clicking on a hyperlink to display a dialog box for saving a document to a local disk instead of displaying it. Either specify the directory and file name or accept the default values. In Chapter 4, you used this option to save special images and movies.

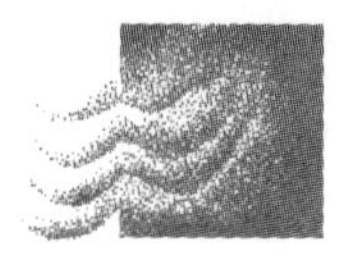

CAUTION

Disable the Load to Local Disk option immediately after loading the desired images to disk. Otherwise, all incoming files will be stored to disk. You won't see documents on the screen and you may run out of disk space.

- **Delay Image Loading** controls whether text and images are transmitted simultaneously. Turning on this option reduces the transmission time for text data.
- **Load Images in Current** shows or suppresses images in a transmitted document. Turn this option off to reduce transmission time, especially in busy networks.
- **Reload Config Files** loads a copy of your previous configuration files. You can use this option to test different Mosaic configurations.
- **Flush Image Cache** removes presently saved (cached) images, to reclaim system memory.
- **Clear Global History** deletes the current history file, which can grow beyond manageable proportions in a system with heavy search demands.
- **Fonts** allows you to select among numerous fonts for different portions of your document.
- **Anchor Underlines** provides a choice (light, medium, heavy, or none at all) of underlining styles for hyperlinks.

The Navigate Menu

The Navigate menu lets you make your way efficiently across the Internet. As you learned in Chapters 4 and 5, you can access this menu repeatedly to get the information you want. The Navigate menu includes the eight options described here:

- **Back** retrieves the most recently accessed document. Use this repeatedly to access documents in reverse order.
- **Forward** is the opposite of the Back option. Use this menu item when you overshoot the desired document.
- **Home Document** returns you to your Home Page.
- **Window History** displays a list of documents in the order you accessed them. Select any item in the list and click on the Load button to obtain the selected document. Click on the Dismiss button to close this window. Click on the Mail button to send this information to NCSA Developers for help in debugging.
- **Hotlist** lets you manipulate a Hotlist of documents for fast reference. Use the AddCurrent button to add a document to this list, the GoTo button to immediately reference this document, the Remove button to delete a document from the Hotlist, the Edit button to change the Hotlist, and the Title button to add a title to the document.
- **Add Current to Hotlist...** places the current document in a special area for documents you wish to retrieve frequently. This is equivalent to clicking on the AddCurrent button in the Hotlist dialog box.
- **Internet Starting Points** specifies where to start searching in the Internet.
- **Internet Resources Meta-Index** brings up a document which is a "loosely" categorized meta-index of the various resource directories and indexes available on the Internet.

The Annotate Menu

You use the Annotate menu to add your comments to Mosaic documents for future reference. The Annotate menu includes the three selections described next.

- **Annotate** displays a special window in which you compose your comments about the given file. *Personal comments* can only be read by you, *group comments* can be read by anyone in your workgroup (provided that the appropriate selection was made in the Show Group Annotations item on the Options menu), and *public comments* can be read by anyone with access to the given document.

- **Audio Annotate...** opens the Audio Annotate window in which you can attach verbal comments to your document, provided that you have installed and configured the proper audio equipment. Clicking on this selection generates a dialog box including Start Recording, Stop Recording, Commit, Dismiss, and Help buttons. The annotation may be personal, workgroup, or public.
- Edit This Annotation enables you to change the contents of the designated annotation.
- **Delete This Annotation** removes the designated annotation, permanently.

The Help Menu

You use the Help menu to obtain help when you need it. One reason Mosaic has become so popular is that its Help screens are easy to track down and easy to use. The Help menu provides the ten selections described next.

- **About...** presents brief information about Unix Mosaic, including the product name, version number, a short copyright notice, the developers' names, and an e-mail address for comments and bug reports.
- **Manual...** provides an online manual; this option is not yet fully implemented.
- **What's New**... describes some of the newest features of Mosaic.
- **Demo...** provides a short demonstration of Mosaic.
- **On Version 2.4...** describes version 2.4 of Mosaic (the current version), comparing it to the previous version.
- **On Window...** explains the components of the NCSA Mosaic screen and menus.
- **On FAQ...** provides answers to frequently asked questions.
- **On HTML...** describes the Mosaic hypertext language
- **On URLs...** describes Universal Resource Locators.
- **Mail Developers...** lets you to compose and send an information request to NCSA developers.

Mosaic Buttons

Sometimes it's quicker to click on a button than to select one or more menu items. The MS-Windows version of Mosaic features a toolbar at the top of the screen that includes 11 icons. The X Window (Unix) version of Mosaic includes nine labeled buttons at the bottom of the screen. Many of the buttons have similar functions in the two versions of Mosaic.

Mosaic Toolbar for Windows

The Mosaic toolbar for Windows contains 12 icons, not all of which are presently implemented. Position the mouse at a toolbar icon to display a short description of it, known as a *tool tip*. While you can turn off the toolbar (by turning off the Show Toolbar item in the Options menu), it's preferable to leave it on. Next you'll learn about the toolbar icons, from left to right.

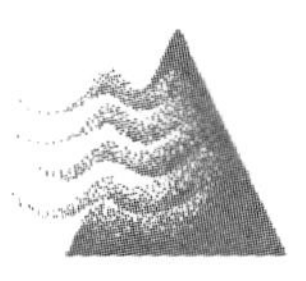

TIP
The Toolbar can be a real time-saver. Use it.

The Open URL Icon

The Open URL icon looks like a file folder, as does the Open Document icon in MS-Word for Windows and other MS-Windows software. Clicking on this icon displays the Open URL dialog box and provides you with access to the Hotlist.

The Save Document URL Icon

The Save Document URL icon looks like a sheet of paper. Clicking on this icon saves the current document to disk.

The Back Icon

The Back icon looks like a left arrow. Clicking on this icon displays the previous document in the history list.

The Forward Icon

The Forward icon resembles a right arrow. Clicking on this icon displays the next document in the history list.

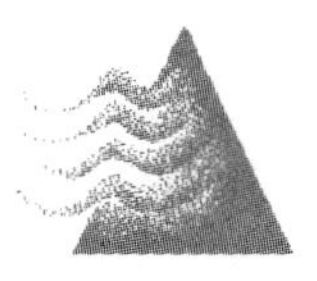

TIP
Click on the Forward icon if you overshoot your mark when attempting to access previous documents.

The Reload Icon

The Reload icon looks like a circle. Clicking on this icon reloads the current document.

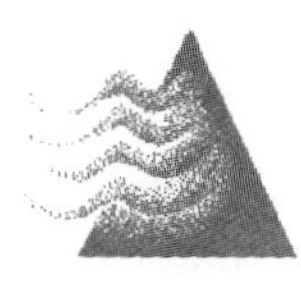

TIP
Click on the Reload icon if you accidentally aborted the document transfer by clicking on the scroll box too soon.

The Home Icon

The Home icon looks like a house. Clicking on this icon generates the default Home Page.

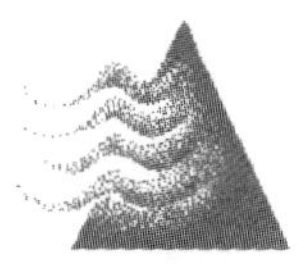

TIP
Click on the Home icon if you have forgotten the specific Starting Points name.

The Add to Hotlist Icon

The Add to Hotlist Icon adds the current document to the Hotlist.

The Copy Icon

The Copy icon contains two overlapping squares. Clicking on this icon copies the current document to the Clipboard.

The Paste Icon

The Paste icon looks like a clipboard. Clicking on this icon pastes the contents of the Clipboard into the active window (but not into the document itself).

The Find Icon

The Find icon resembles a document with a downward pointing arrow. Clicking on this icon displays the Find dialog box for locating a specified character string within the document.

The Print Icon

The Print icon looks like a printer. Clicking on this icon prints the current document.

The About Mosaic Icon

The About Mosaic icon looks like a question mark. Clicking on this icon displays basic information about Mosaic.

Mosaic Buttons for X Window (Unix)

Mosaic for X Window contains nine buttons at the bottom of the screen. Now you'll learn about these buttons, from left to right.

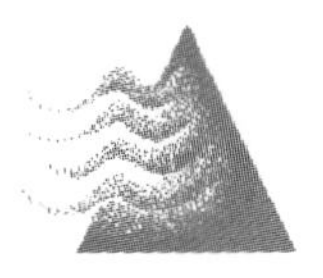

TIP
These buttons can be real time-savers. Use them.

The Back Button

Clicking on the Back button displays the previous document in the history list.

The Forward Button

Clicking on the Forward button displays the next document in the history list.

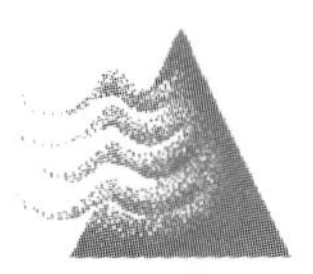

TIP
Click on the Forward button if you overshoot your mark when attempting to access previous documents.

The Home Button

Clicking on the Home button generates the default Home Page.

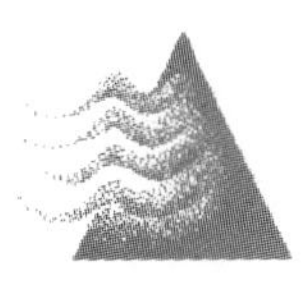

TIP
Click on the Home button if you have forgotten the specific Starting Points name.

The Reload Button

Clicking on the Reload button reloads the present document.

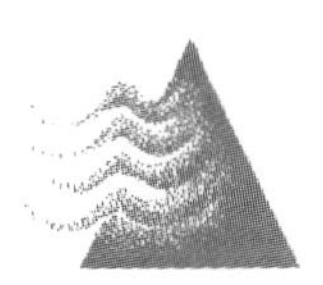

TIP
Click on the Reload button if you accidentally aborted the document transfer by clicking on the scroll box too soon.

The Open Button

Clicking on the Open button displays the Open URL dialog box and provides you with access to the Hotlist.

The Save As Button

Clicking on the Save As button saves the current document to disk.

The Clone Button

Clicking on the Clone button creates an additional copy of the current document window.

The New Window Button

Clicking on the New Window button creates a new, empty window.

The Close Window Button

Clicking on the Close Window button closes the current window.

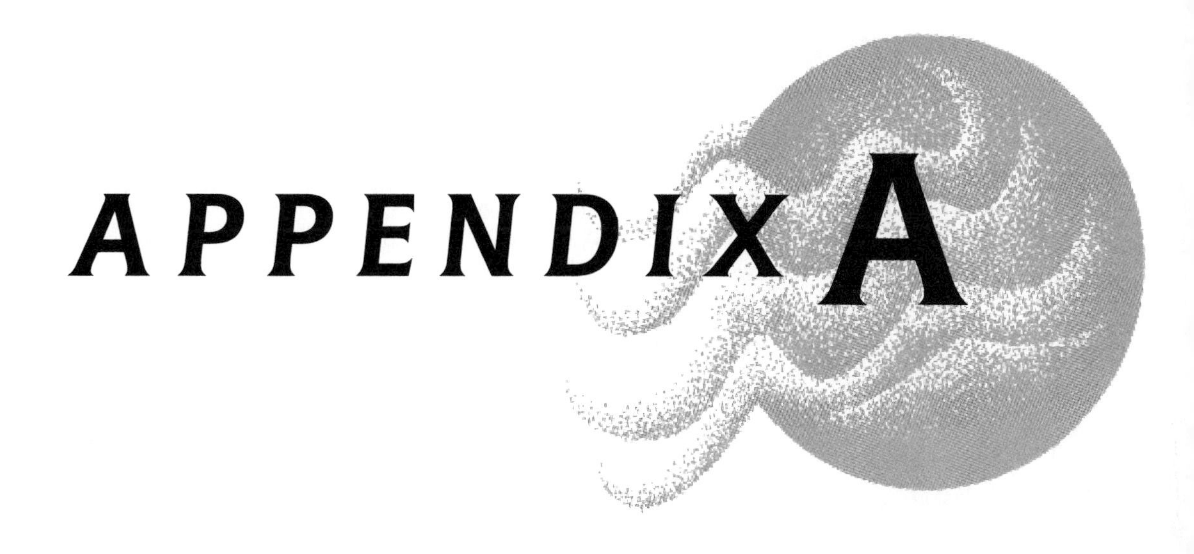

Useful Information Sources

This appendix is composed of two parts. The first part is a compendium of all information sources that appeared within this text. The second part is a list of useful Gopher Servers. Both parts include a short reminder of how to access them from Mosaic.

Compendium of Information Sources Appearing in Text

This compendium is organized in alphabetical order by chapter and includes both information sources referred to within the text and those appearing in the figures. Chapter 5 showed you how to enter URLs for Servers. To do so, access the Open URL option of the File menu to generate the Open URL dialog box. Then enter the URL as requested. For example, to access the World Cup USA '94 Server, enter **http://www.sun.com/wc94/**.

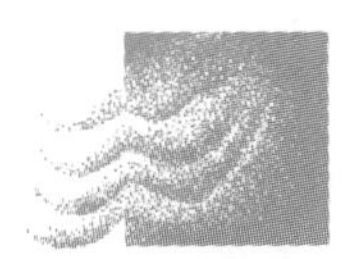

CAUTION
Because of typesetting constraints, some URLs appear here on two lines. They must be entered on a single line.

Chapter 1

INFORMATION SOURCE	URL
Britannica Online	**http://www.eb.com**
Canadian Weather Forecast	**gopher://wx.atmos.uiuc.edu/11/Canada**
CBC Radio Trial	**http://debra.dgbt.doc.ca/cbc/cbc.html**
CommerceNet Home	**http://www.commerce.net/**
COSTA Table of Contents	**http://mmink.cts.com/mmink/kiosks/costa/ costatravel.html**
Digital World Wide Web Server	**http://www.digital.com/**
HomeBuyer's Fair Welcome	**http://www.homefair.com/homepage.html**
Intel	**http://www.intel.com/**
Novell Inc. World Wide Web Home Page	**http://www.novell.com/**
Overview of the Web	**http://info.cern.ch/hypertext/WWW/LineMode/ Defaults/default.html**
SCO World Wide Web Home Page	**http://www.sco.com/**
World Cup USA '94	**http://www.sun.com/wc94/**
World Wide Web Servers	**http://www.eit.com/web/www.servers/ www.servers.html**

Chapter 2

INFORMATION SOURCE	URL
NCSA anonymous ftp transfer address	**ftp://ftp.ncsa.uiuc.edu**
NCSA Mosaic for Microsoft Windows User's Guide	**http://www/ncsa.uiuc.edu/SDG/Software/WinMosaic/Docs/WMosTOC.html**
NCSA Mosaic Home Page	**http://www/ncsa.uiuc.edu/SDG/Software/Mosaic/NCSAMosaicHome.html**
World Wide Web—The Project	**http://info.cern.ch/hypertext/WWW/TheProject.html**

Chapter 3

INFORMATION SOURCE	URL
Internet Resource Meta-Index	**http://www.ncsa.uiuc.edu/SDG/Software/Mosaic/Meta/ndx.html**
NCSA Mosaic Home Page	**http://www.ncsa.uiuc.edu/SDG/Software/Mosaic/NCSAMosaicHome.html**
Starting Points for Internet Exploration	**http://www.ncsa.uiuc.edu/SDG/Software/Mosaic/StartingPoints/NetworkStartingPoints.html**

Chapter 4

INFORMATION SOURCE	URL
19th Hole	**http://dallas.nmhu.edu/golf/golf.htm**
Carnegie Mellon University SCS Front Door Page	**http://www.cs.cmu.edu:8001/Web/FrontDoor.html**
Collision of Jupiter and Comet Shoemaker-Levy 9	**http://pscinfo.psc.edu/research/user_research/mac_low/mac_low.html**

Comet P/Shoemaker-Levy 9 Impact Home Page	**http://seds.lpl.arizona.edu/sl9/sl9.html**
Commercial World Wide Web Servers	http://www.eit.com/web/www.servers/commercial.html
English Server	**http://english-server.hss.cmu.edu/**
GolfData On-Line Home Page	**http://www.gdol.com/**
Honululu Community College WWW Service	**http://www.hcc.hawaii.edu/**
Known Golf Links	**http://www.gdol.com/golf.links.html**
On the day of June 20th	**gopher://info.tamu.edu:70/00/.data/politics/1994/sched.0620**
Radio Stations KKSF and KDFC	**http://kksf.tbo.com/**
What's New With NCSA Mosaic	**http://www.ncsa.uiuc.edu/SDG/Software/Mosaic/Docs/whats-new.html**
White House	**http://english-server.hss.cmu/WhiteHouse.httml**

Chapter 5

INFORMATION SOURCE	URL
BART System Map	**http://server.berkeley.edu/Transit/BART/BART_System.html**
Bay Area Restaurant Guide	**http://netmedia.com/ims/rest/ba_rest_guide.html**
California World Wide Web Servers	**http://www.linl.gov/ptolls/california.servers.html**
Canadian Weather Forecast	**gopher://wx.atmos.uiuc.edu/11/Canada**
CERN Virtual Library	**http://info.cern.ch/hypertext/DataSources/bySubject/Overview.html"**
Digital Multivendor Customer Services	**http://www.service.digital.com/**
Digital Performance Trouble Report	**http://www.digital.com/hypertext/util/pathcheck.html**
Interactive BART Fare Schedule	**http://server.berkeley.edu/Transit/BART/BartMap.cgi/N=1&Sta1=LAKEM&Sta2=?216,151**

Internet Newsgroups	**http://info.cern.ch/hypertext/DataSources/News/Groups/Overview.html**
NCSA Mosaic alternative newsgroups	**http://info.cern.ch/hypertext/DataSources/News/Groups/alt.html**
News with Mosaic	**http://www.ncsa.uiuc.edu/SDG/Software/Mosaic/Docs/whats-new.html**
Newsgroups	**alt.bbs news:alt.bbs**
Searchable Gopher Index	**gopher://veronica.uni-koeln.d:2347/7**
Starting Points for Internet Exploration	**http://www.ncsa.uiuc.edu/SDG/Software/Mosaic/StartingPoints/NetworkStartingPoints.html**
The WWW Virtual Library Subject Catalogue	**http://info.cern.ch/hypertext/DataSources/bySubject/Overview.html**
Travelers Forecast for the Major Cities	**gopher://wx.atmos.uiuc.edu:70/00/Regional/Travelers%20Forecast%20Table%20%2810%29**
University of Illinois at Urbana-Champaign	**http//www.uiuc.edu/**
Veronica Search Results	**gopher://veronica.uni-koeln.d:2347/7?Open+Computing**
Veronica Search Results	**gopher://veronica.uni-koeln.d:2347/7?Networking**
Virtual Tourist—North America	**http://wings.buffalo.edu/world/na.html**
Virtual Tourist—World	**http://wings.buffalo.edu/world**
WAIS Search Results	**http://info.cern.ch.8001/quake.think.com/directory-of-servers?www+and+mosaic**
WAIS Sources	**http://info.cern.ch/hypertext/Products/WAIS/Sources/Overview.html**
WWW Info	**http://wings.buffalo.edu/world**
WWW Servers	**http://www.eit.com/web/www.servers/www.servers.html**
X Window Mosaic Documentation	**http://www.ncsa.uicu.edu/SDG/Software/Mosaic/Docs/UserGuide /Xmosaic.0.html**

Chapter 6

INFORMATION SOURCE	URL
World Wide Web	**http://info.cern.ch/hypertext/WWW/TheProject.html**

Chapter 7

INFORMATION SOURCE	URL
Bug list	**http://www.ncsa.uiuc.edu/SDG/Software/WinMosaic/Bugs.html**
FAQ Page	**http://www.ncsa.uiuc.edu/SDG/Software/WinMosaic/FAQ.html**
Feature Page	**http://www.ncsa.uiuc.edu/SDG/Software/WinMosaic/Features.html**
Finger Gateway	**http://cs.indiana.edu/finger/gateway**
Help Page	**http://www.ncsa.uiuc.edu/SDG/Software/WinMosaic/Docs/WMosTOC.html**
NCSA Mosaic Demo Document	**http://www.ncsa.uiuc.edu/demoweb/demo.html**
NCSA Mosaic's "What's New" Page	**http://www.ncsa.uiuc.edu/SDG/Software/Mosaic/Docs/whats-new.html**
Starting Points Document	**http://www.ncsa.uiuc.edu/SDG/Software/Mosaic/StartingPoints/NetworkStartingPoints.html**
Whois Gateway	**gopher://sipb.mit.edu:70/1B%3aInternet%20whois%20servers**

Appendix B

INFORMATION SOURCE	URL
ftp Test File	**//ftp.ncsa.uiuc.edu/PC/Mosaic/faq.txt**
http Test File	**http://www.ncsa.uiuc.edu/General/UIUC/UIUCIntro/UofI_intro.html**
Viewers that run with Mosaic	**http://www.ncsa.uiuc.edu/SDG/Software/WinMosaic/viewers.html**

Gopher Servers

Chapter 5 showed you how to enter URLs for Gopher Servers. To do so, access the Open URL option of the File menu to generate the Open URL dialog box. Then enter the URL as requested. For example, to access the Academe This Week (Chronicle of Higher Education) Gopher Server, enter **gopher://chronicle.merit.edu**.

GOPHER SERVER	URL
Academe This Week (Chronicle of Higher Education)	**gopher://chronicle.merit.edu**
Anesthesiology Gopher	**gopher://eja.anes.hscsyr.edu**
Apple Computer Higher Education Gopher Server	**gopher://info.hed.apple.com**
Arabidopsis Research Companion, Mass Gen Hospital/Harvard	**gopher://weeds.mgh.harvard.edu**
AskERIC—(Educational Resources Information Center)	**gopher://ericir.syr.edu**
Austin Hospital, Melbourne, Australia	**gopher://pet1.austin.unimelb.edu.au**
Australian Defence Force Academy, Canberra, Australia	**gopher://gopher.adfa.oz.au**
Australian Environmental Resources Information Network (ERIN)	**gopher://kaos.erin.gov.au**
Baylor College of Medicine	**gopher://gopher.bcm.tmc.edu**
Bedford Institute of Oceanography, Canada	**gopher://biome.bio.dfo.ca**
Biodiversity and Biological Collections Gopher	**gopher://huh.harvard.edu**
Bioftp EMBnet, CH	**gopher://bioftp.unibas.ch**
BioInformatics gopher at ANU	**gopher://life.anu.edu.au**
CAMIS (Center for Advanced Medical Informatics at Stanford)	**gopher://camis.stanford.edu**
Centre for Scientific Computing, FI	**gopher://gopher.csc.fi**
CICNET gopher server	**gopher://nic.cic.net**
CIESIN Global Change Information Gateway	**gopher://gopher.ciesin.org**
Cleveland State University Law Library	**gopher://gopher.law.csuohio.edu**

Colorado State University Optical Computing Lab	**gopher://sylvia.lance.colostate.edu**
Computational Biology (Welchlab - Johns Hopkins University)	**gopher://merlot.welch.jhu.edu**
Computer Solutions by Hawkinson	**gopher://csbh.com**
Consortium for School Networking (CoSN)	**gopher://cosn.org**
Cornell Law School, experimental	**gopher://fatty.law.cornell.edu**
Cornell Medical College	**gopher://gopher.med.cornell.edu**
CPSR (Computer Professionals for Social Responsibility)	**gopher://gopher.cpsr.org**
CREN/Educom	**gopher://info.educom.edu**
CWRU Medical School—Department of Biochemistry	**gopher://biochemistry.cwru.edu**
CYFER-net USDA ES Gopher Server	**gopher://cyfer.esusda.gov**
Dana-Farber Cancer Institute, Boston, MA	**gopher://gopher.dfci.harvard.edu**
Dendrome: Forest Tree Genome Mapping Database	**gopher://s27w007.pswfs.gov**
Department of Information Resources (State of Texas), experimental	**gopher://ocs.dir.texas.gov**
DNA Data Bank of Japan, Natl. Inst. of Genetics, Mishima	**gopher://gopher.nig.ac.jp**
Ecogopher at the University of Virginia	**gopher://ecosys.drdr.virginia.edu**
EDUCOM Documents and News	**gopher://educom.edu**
Electronic Frontier Foundation	**gopher://gopher.eff.org**
Envirogopher (at CMU)	**gopher://envirolink.hss.cmu.edu**
Extension Service USDA Information	**gopher://zeus.esusda.gov**
Federal Info. Exchange (FEDIX), experimental	**gopher://fedix.fie.com**
Genethon (Human Genome Res. Center, Paris), FR	**gopher://gopher.genethon.fr**
GrainGenes, the Triticeae Genome Gopher	**gopher://greengenes.cit.cornell.edu**
GRIN, National Genetic Resources Program, USDA-ARS	**gopher://gopher.ars-grin.gov**

Gustavus Adolphus College	**gopher://gopher.gac.edu**
HENSA micros (National software archive, Lancaster univ.), UK	**gopher://micros.hensa.ac.uk**
Human Genome Mapping Project Gopher Service, UK	**gopher://menu.crc.ac.uk**
ICGEBnet, Int.Center for Genetic Eng. & Biotech, IT	**gopher://genes.icgeb.trieste.it**
ICTP, International Centre for Theoretical Physics, Trieste, IT	**gopher://gopher.ictp.trieste.it**
Info Mac Archives (sumex-aim)	**gopher://SUMEX-AIM.Stanford.EDU**
INN, Weizmann Institute of Science, Israel	**gopher://sunbcd.weizmann.ac.il**
Institute of Physics, University of Zagreb, HR	**gopher://gopher.ifs.hr**
InterCon Systems Corporation	**gopher://vector.intercon.com**
Internet Society	**gopher://ietf.CNRI.Reston.Va.US**
Internet Wiretap	**gopher://wiretap.spies.com**
InterNIC—Internet Network Information Center	**gopher://rs.internic.net**
ISU College of Pharmacy	**gopher://pharmacy.isu.edu**
IUBio Biology Archive, Indiana University, experimental	**gopher://ftp.bio.indiana.edu**
IUPUI Integrated Technologies	**gopher://indycms.iupui.edu**
JvNCnet	**gopher://gopher.jvnc.net**
LANL Physics Information Service	**gopher://mentor.lanl.gov**
Library of Congress (LC MARVEL)	**gopher://marvel.loc.gov**
Library X at Johnson Space Center	**gopher://krakatoa.jsc.nasa.gov**
Liverpool University, Dept of Computer Science, UK	**gopher://gopher.csc.liv.ac.uk**
Los Alamos National Laboratory	**gopher://gopher.lanl.gov**
Maize Genome Database Gopher	**gopher://teosinte.agron.missouri.edu**
McGill Research Centre for Intelligent Machines, Montreal, Canada	**gopher://lightning.mcrcim.mcgill.edu**
Merit Network	**gopher://nic.merit.edu**
Michigan State University	**gopher://gopher.msu.edu**

NASA Goddard Space Flight Center	**gopher://gopher.gsfc.nasa.gov**
NASA Mid-Continent Technology Transfer Center	**gopher://technology.com**
NASA Network Applications and Information Center (NAIC)	**gopher://naic.nasa.gov**
NASA Shuttle Small Payloads Info	**gopher://vx740.gsfc.nasa.gov**
National Cancer Center, Tokyo Japan	**gopher://gopher.ncc.go.jp**
National Center for Supercomputing Applications	**gopher://gopher.ncsa.uiuc.edu**
National Center on Adult Literacy	**gopher://litserver.literacy.upenn.edu**
National Institute of Allergy and Infectious Disease (NIAID)	**gopher://gopher.niaid.nih.gov**
National Institute of Standards and Technology (NIST)	**gopher://gopher-server.nist.gov**
National Institutes of Health (NIH) Gopher	**gopher://gopher.nih.gov**
National Science Foundation Gopher (STIS)	**gopher://stis.nsf.gov**
North Carolina State University Library Gopher	**gopher://dewey.lib.ncsu.edu**
Nova Scotia Technology Network, N.S., Canada	**gopher://nstn.ns.ca**
Novell Netwire Archives	**gopher://ns.novell.com**
Oak Ridge National Laboratory ESD Gopher	**gopher://jupiter.esd.ornl.gov**
ONENET Networking Information	**gopher://nic.onenet.net**
Oregon State University Biological Computing (BCC)	**gopher://gopher.bcc.orst.edu**
PeachNet Information Service	**gopher://gopher.PeachNet.EDU**
Physics Resources, experimental	**gopher://granta.uchicago.edu**
PIR Archive, University of Houston	**gopher://ftp.bchs.uh.edu**
Presbyterian College (Clinton, SC)	**gopher://cs1.presby.edu**
Primate Info Net (University of Wisconsin-Madison)	**gopher://saimiri.primate.wisc.edu**
PSGnet/RAINet—low-cost and international networking	**gopher://gopher.psg.com**

Saint Louis University	**gopher://sluava.slu.edu**
Scholarly Communications Project Electronic Journals	**gopher://borg.lib.vt.edu**
South African Bibliographic and Information Network	**gopher://info2.sabinet.co.za**
Soybean Data	**gopher://mendel.agron.iastate.edu**
Space Telescope Electronic Information System (STEIS)	**gopher://stsci.edu**
Sprintlink gopher Server, Virginia USA	**gopher://ftp.sprintlink.net**
Stanford University Medical Center	**gopher://med-gopher.stanford.edu**
State University of New York (SUNY)—Brooklyn Health Science Center	**gopher://gopher1.medlib.hscbklyn.edu**
State University of New York (SUNY)—Syracuse Health Science Center	**gopher://micro.ec.hscsyr.edu**
TECHNET, Singapore	**gopher://solomon.technet.sg**
Texas A&M	**gopher://gopher.tamu.edu**
Texas Internet Consulting (TIC), Austin, TX	**gopher://gopher.tic.com**
The World (Public Access UNIX)	**gopher://world.std.com**
Universite de Montreal Megagopher	**gopher://megasun.bch.umontreal.ca**
University of California—Davis, CA	**gopher://gopher.ucdavis.edu**
University of California—Irvine, CA	**gopher://gopher-server.cwis.uci.edu**
University of California—Santa Barbara Geological Sciences, CA	**gopher://gopher.geol.ucsb.edu**
University of Chicago Law School	**gopher://lawnext.uchicago.edu.**
University of Georgia, GA	**gopher://gopher.uga.edu**
University of Houston Protein Information Resource, TX	**gopher://ftp.bchs.uh.edu**
University of Illinois at Chicago, IL	**gopher://gopher.uic.edu**
University of Illinois at Urbana-Champaign, IL	**gopher://gopher.uiuc.edu**
University of Michigan Libraries, MI	**gopher://gopher.lib.umich.edu**
University of Minnesota Soil Science Gopher Information Service	**gopher://gopher.soils.umn.edu**
University of Nevada	**gopher://gopher.unr.edu**

University of New Mexico	**gopher://peterpan.unm.edu**
University of Notre Dame	**gopher://gopher.nd.edu**
University of Texas at El Paso, Geological Sciences Department	**gopher://dillon.geo.ep.utexas.edu**
University of Texas Health Science Center at Houston	**gopher://gopher.uth.tmc.edu**
University of Texas M. D. Anderson Cancer Center	**gopher://utmdacc.uth.tmc.edu**
University of Texas Medical Branch	**gopher://phil.utmb.edu**
University of Washington, Pathology Department	**gopher://larry.pathology.washington.edu**
University of Wisconsin—Madison, Medical School	**gopher://msd.medsch.wisc.edu**
US Geological Survey (USGS)	**gopher://info.er.usgs.gov**
USCGopher (University of Southern California)	**gopher://cwis.usc.edu**
USGS Atlantic Marine Geology	**gopher://bramble.er.usgs.gov**
Vertebrate World Server at Colorado State University	**gopher://neptune.rrb.colostate.edu**
Virginia Coast Reserve Information System (VCRIS)	**gopher://atlantic.evsc.Virginia.EDU**
Vortex Technology	**gopher://gopher.vortex.com**
Whole Earth 'Lectronic Magazine—The WELL's Gopherspace	**gopher://gopher.well.sf.ca.us**
Worcester Foundation for Experimental Biology	**gopher://sci.wfeb.edu**
World Data Center on Microorganisms (WDC), Riken, Japan	**gopher://fragrans.riken.go.jp** **gopher://gopher.mobot.org**
World Health Organization (WHO)	**gopher://gopher.who.ch**
Yale University	**gopher://yaleinfo.yale.edu**

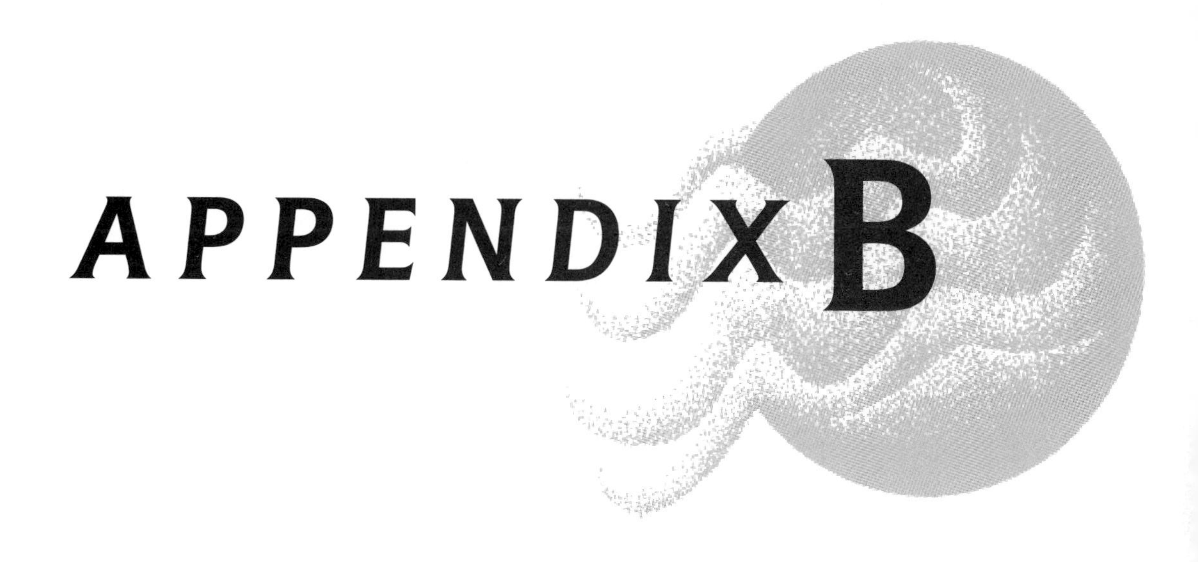

Installing Mosaic

This appendix tells you what you need to know to install Mosaic. However, before you can install Mosaic you may have to install software that enables you to connect to the Internet by using the TCP/IP protocol. We provide the details about two widely used communications software packages for the Internet environment: Chameleon and Trumpet. Then we take a brief look at HookUp Communications, one of the many Internet providers on the market. Next we examine how to install the two versions of Mosaic covered in this book. As you would expect, the procedures for installing and running Mosaic differ somewhat for the MS-Windows and Unix versions.

Before you start installing software, let's review the two types of connections to the Internet. A direct connection occurs when you are part of a network connected to the Internet. In this case, you don't need SLIP or PPP but you do need TCP/IP. If you don't have a direct Internet connection—for example, if you are trying to run

Mosaic from a stand-alone computer—you need an Internet provider to run SLIP or PPP. In both cases, you need to install a communications package.

WIN
Two widely used communications packages are Chameleon, a supported commercial product, and Trumpet, a shareware product. Both products are discussed in this appendix.

X
TCP/IP is part of Unix. You must have a privileged account (defined in Unix as "root") to configure it. Different versions involve different menu-driven or GUI setups. The manual setup procedures will require more advanced Unix experience. See *Unix System Administration Guide*, by Levi Reiss and Joseph Radin (Osborne/McGraw-Hill, 1993).

Using Internet Chameleon

Internet Chameleon, often called Chameleon for short, is one of the leading TCP/IP connectivity products for MS-Windows. It was designed specifically for MS-Windows. It allows you to connect to the Internet as a full TCP/IP host, with your own IP address and host name. Chameleon is preconfigured for many Internet providers, including HookUp Communications, which we used to access the Internet.

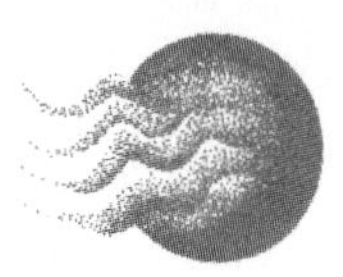

NOTE
Chameleon is designed for mobile and home users who need dial-up access to the Internet.

Chameleon features include the following:

- Multiple windows running multiple TCP/IP sessions
- Applications running both client and server functions
- A Windows point-and-click interface
- Context-sensitive online help
- SLIP and PPP interfaces

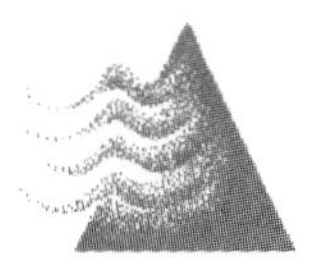

TIP
More information about Chameleon can be obtained from the supplier, NetManage at 10725 North De Anza Blvd., Cupertino, CA 95014. Their telephone is (408) 973-7171, their fax is (408) 257-6405, and their e-mail address is sales@netmanage.com.

Requirements Checklist

Chameleon requires the following minimum hardware configuration: an IBM-compatible PC with a 386 or later processor, a hard disk with about 5MB of free space, 4MB or more of memory, and a modem that runs from a COM port. The software requires version 5.0 or later of DOS, and MS-Windows 3.1 or later, running enhanced mode.

You must have an account with an Internet provider for SLIP or PPP dialup access. The following information is required and your Internet provider will supply the values.

- The Interface type (such as *PPP*)
- An Internet address (such as *154.14.1.16*)
- The host name (such as *jradin*)
- The domain name (such as *hookup.net*)
- The domain name server address (such as *154.14.1.1*)
- A telephone number (such as *484-1256*)
- A user name (such as *jradin*)
- A user password (such as *thepassword*)

You are responsible for the following settings associated with your modem.

- The COM port (such as *COM2*)
- The baud rate (such as *14,400*)
- The modem type (such as *Hayes*)

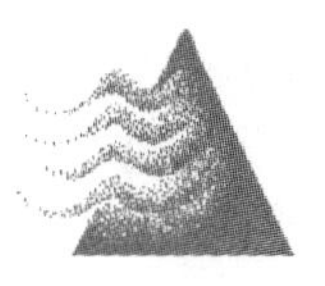

TIP
Don't use a modem whose baud rate is less than 9,600, because the information transfer will be painfully slow.

In addition to providing a connection to the Internet, Chameleon lets you use mail and news. In this case, the Internet provider will supply the information:

- The mail server name (such as *pop.hookup.net*)
- The mail gateway name (such as *mail.hookup.net*)
- The Internet mailbox name (such as *jradin*)
- The Internet mailbox password (such as *thepassword*)
- News server address (such as *147.157.2.1*)

Installation

Chameleon comes on three floppy disks. The setup disk contains the setup and application program. The remaining disks contain the TCP/IP kernel and network card drivers. To install the program, place the setup disk in drive A, type **a:setup**, and follow the instructions.

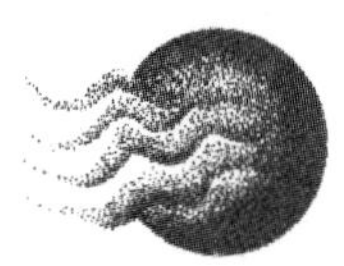

NOTE
The installation also works from drive B.

The installation creates an Internet Chameleon program group, as shown in Figure B-1.

Chameleon creates a new directory, **c:\netmanag** by default, and appends two lines such as the following to your **win.ini** file.

```
[tcpip]
file:c:\netmanag\tcpip.cfg
```

The following line is added to the beginning of the *Path* variable in the **autoexec.bat** file.

```
set path=c:\netmanag;
```

Using Dial-up Connections

Chameleon supplies scripts for connecting to many Internet providers, including HookUp Communications, the Internet provider that furnished our PPP account.

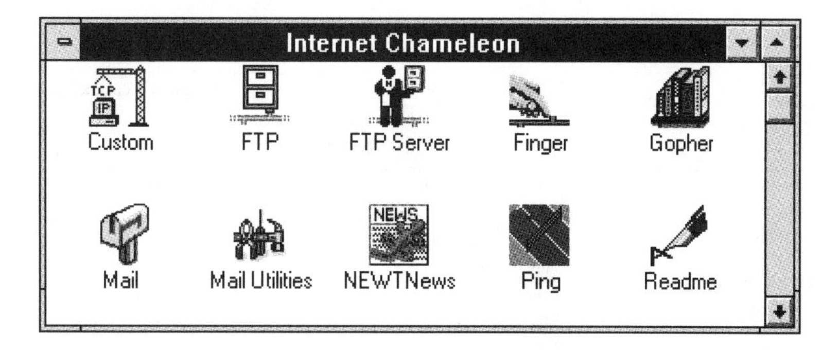

FIGURE B-1. *The Internet Chameleon program group*

First you install Chameleon and start the Custom application by double-clicking on its icon. Then you choose Open from the File menu and select the file name associated with your Internet provider.

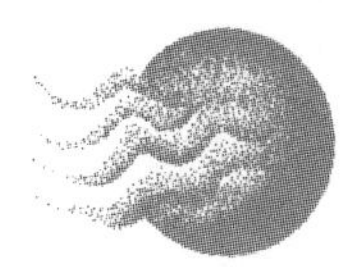

NOTE
If the Open Configuration File window, shown in part in Figure B-2, does not contain a file for your Internet provider, you have to complete the entry by typing the appropriate file name in the File Name box.

From the Custom screen, click on the Setup menu item. The ensuing pull-down menu includes several options, such as IP Address and Host Name. Select each in turn, and complete the interface configuration by entering the appropriate values, often supplied by your Internet provider. For example, click on IP Address and enter your specific address in the dialog box. Some settings are shown in Figures B-3 (Port), B-4 (Modem), B-5 (Login), and B-6 (Final Settings).

After completing the interface configuration, you must connect your computer to the Internet via the Custom application. Click on the Connect menu item to launch the network connection process, as shown in Figure B-7.

Accessing the FTP Application

The FTP application invokes the ftp program that transfers files between computers on a network. Chameleon includes both client and server functions.

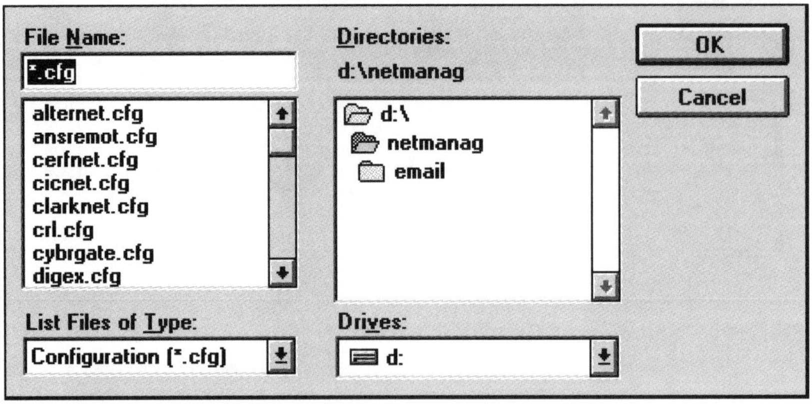

FIGURE B-2. *Open Configuration File window*

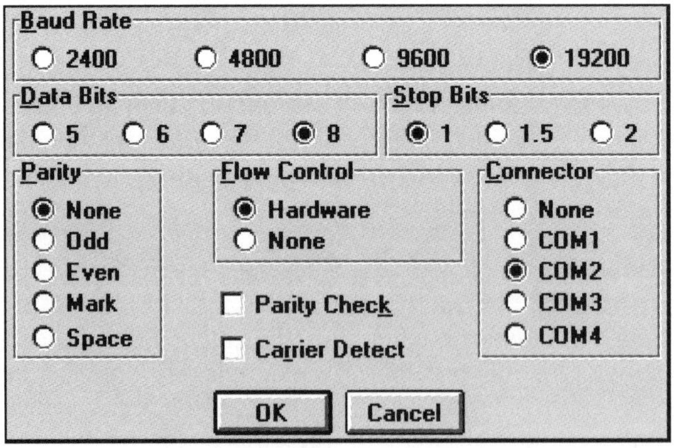

FIGURE B-3. *Port Settings dialog box*

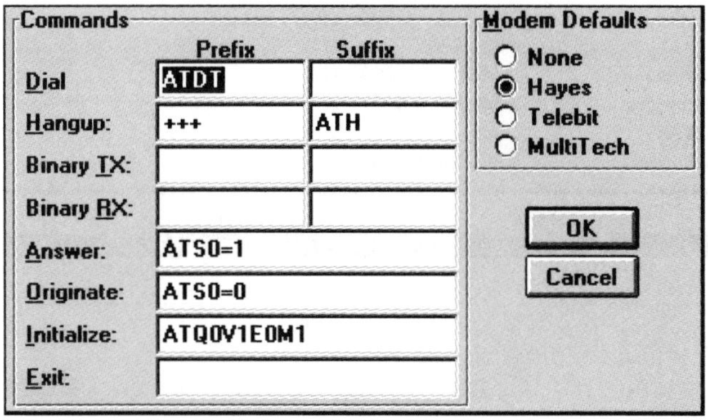

Commands

	Prefix	Suffix
Dial	ATDT	
Hangup:	+++	ATH
Binary TX:		
Binary RX:		
Answer:	ATS0=1	
Originate:	ATS0=0	
Initialize:	ATQ0V1E0M1	
Exit:		

Modem Defaults

- None
- Hayes
- Telebit
- MultiTech

OK

Cancel

FIGURE B-4. *Modem Settings dialog box*

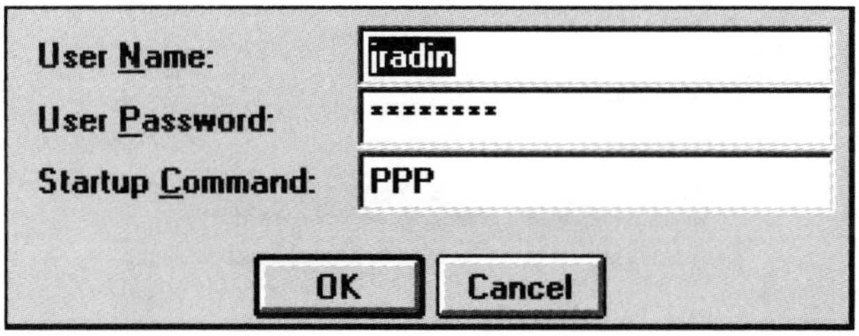

User Name: jradin

User Password: ********

Startup Command: PPP

OK Cancel

FIGURE B-5. *Login Settings dialog box*

Interface:	HookUp - COM2, 19200 baud
Dial:	5265636
IP Address:	165.154.16.160
Subnet Mask:	255.255.0.0
Host Name:	jradin
Domain Name:	ott.hookup.net

Name	Type	IP	Domain
HookUp	PPP	165.154.16.160	ott.hookup.net

FIGURE B-6. *Final Settings dialog box*

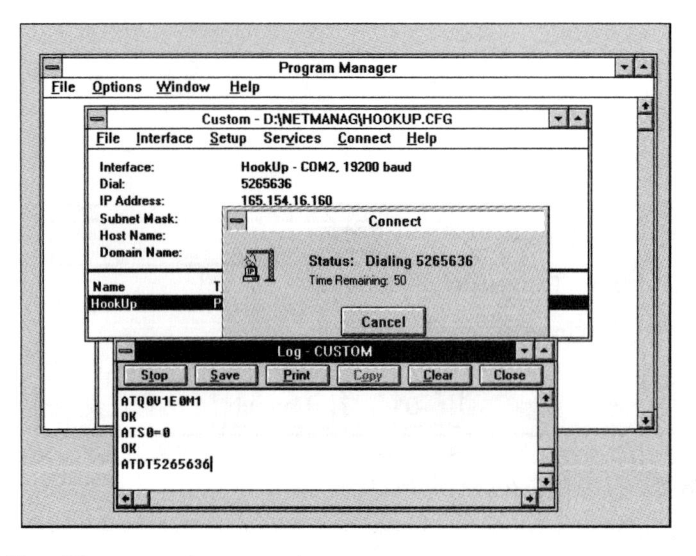

FIGURE B-7. *The network connection process*

Use the following procedure to access a given remote system:

1. Click on the FTP icon to display the FTP screen.
2. Select the Connect menu to display the Connect dialog box.
3. Enter the following host name and enter the user name as anonymous. Then enter the password as *username@address,* as shown here:

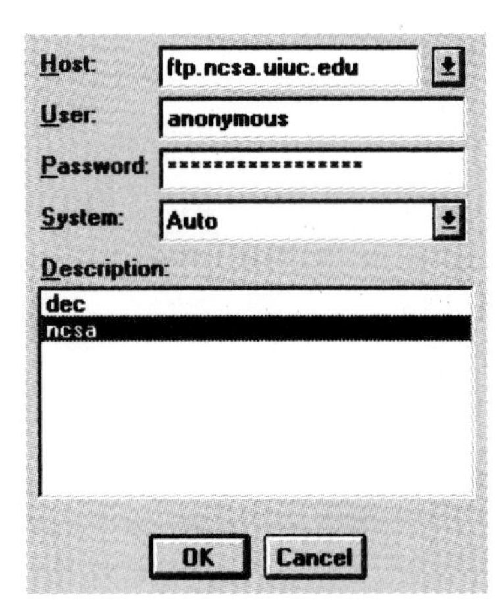

4. Click on the System drop-down list and select the system type, such as Unix, from the ensuing list. The Auto selection will often make the correct choice for you.
5. Confirm your selections by clicking on the OK button.
6. Select the file name on the remote system and then click on Copy to Local with a predefined directory to launch the transfer process, as shown in Figure B-8.

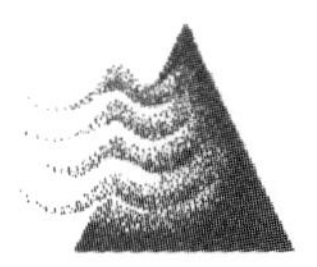

TIP

Select the Settings¦Connection Profile option of the ftp dialog box. Click on New and enter the Description (for example, ncsa), and the remaining parameters such as Host, Remote Dir (where to obtain files), and Local Dir (where to store files), and click on the Close button to confirm. The next time you connect, the appropriate name will appear in the Description list and you need only click to select it (and click on OK to confirm.)

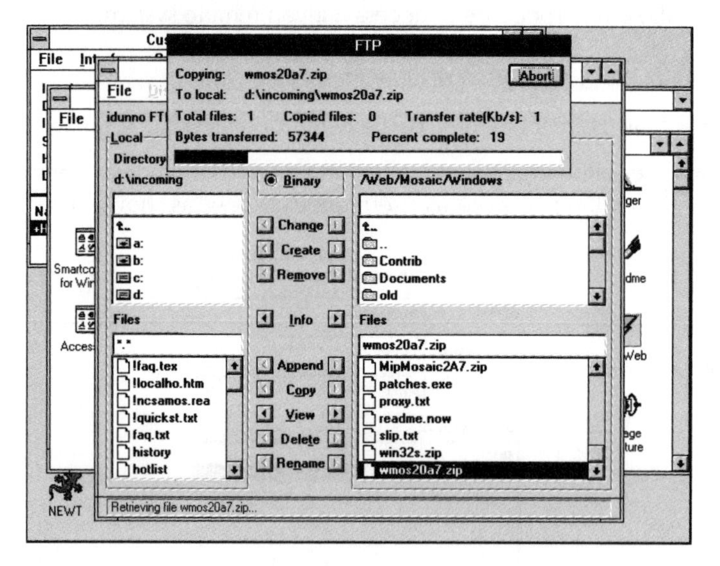

FIGURE B-8. *Making the ftp Connection*

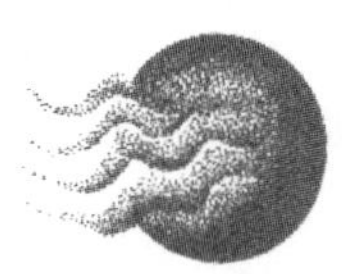

NOTE
The connect pararmeters and options can be saved in a configuration file, named **ftp.cfg** by default. This configuration file is loaded when an ftp application starts.

Trumpet Winsock

Trumpet Winsock is a Windows Sockets 1.1-compatible TCP/IP protocol, providing a standard networking layer for Windows applications such as Mosaic. A compressed copy is found on NCSA's FTP server in the directory **PC/Mosaic/sockets/**. Download the files **disclaim.txt** and **winsock.zip**. This shareware is also found at the anonymous FTP server **ftp.utas.edu.au** in the directory **/pc/trumpet/winsock**.

This section discusses how to install Trumpet Winsock for a SLIP connection. If you are using a direct connection, you should also install a *packet driver,* a software interface between your network card and the TCP program.

Installation

Use the following steps to install Trumpet Winsock:

1. Create a directory such as **c:\trumpet** and copy the following files to the directory:
   - **WINSOCK.DLL** The heart of the TCP/IP driver
   - **TCPMAN.EXE** An interface for setting up Winsock hosts
   - **PROTOCOL** A list of Internet protocols
   - **SERVICES** A list of Internet services
2. Modify the path statement in the **autoexec.bat** file to include the following line:

```
set c:\>set path=c:\trumpet;
```

3. Reboot the system to activate the modified **autoexec.bat** file.

Configuration

Use the following steps to configure Trumpet Winsock:

1. Start MS-Windows.
2. Select Run from the File menu. Type **tcpman** in the dialog box and press the ENTER key to confirm.
3. Click on Internal SLIP.
4. Complete the Setup screen with the following information:
   - **IP ADDRESS** Enter your Internet IP address.

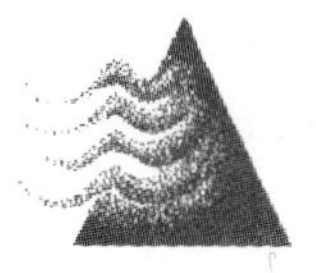

TIP
If you are using a dial-in script that extracts your IP address, you can enter the default address of 0.0.0.0.

   - **NAME SERVER** Specify your name server address for the Domain Name Service (DNS), which converts network names to addresses. Separate multiple addresses with spaces.

- **TIME SERVER** Unused.
- **DOMAIN SUFFIX** A list of domain suffixes, separated by spaces, that you use when resolving names in the Domain Name Service.
- **MTU (Maximum Transmission Unit)** Usually TCP MSS + 40. (TCP MSS is defined in a moment.)
- **TCP RWIN (TCP Receive Window)** Set to about three to four times the value of TCP MSS.
- **TCP MSS (TCP Maximum Segment Size)** Set to 512 bytes for SLIP.
- **SLIP port** Set to 1 for COM1, 2 for COM2, and so forth.
- **BAUD RATE** Set the desired transmission rate, such as 14,400.
- **HARDWARE HANDSHAKE** Recommended if supported by your system.
- **VAN JACOBSON CSLIP COMPRESSION** You may also have to adjust MTU, MSS, & RWIN to be suitable.
- **ONLINE STATUS DETECTION** If supported, select DCD (Data Carrier Detect) or DSR (Data Set Ready) online status detection.

Logging into the Server

Use the following steps to log into the server:

1. Access File Manager in Windows and select the directory where Trumpet was installed.
2. Activate **tcpman.exe** by double-clicking.
3. Log into the server.
4. Press the ESC key after you have finished logging in.
5. If your IP address is allocated dynamically, you may need to set it.
6. Test your connection by executing the **pingw** command, the Trumpet version of the **ping** program which tests network connections.

Internet Provider Setup Program

Internet providers may supply connection software. For example, the Hookup Communications package is based on Trumpet Winsock software. Figure B-9 shows an example of the setup. Figure B-10 shows the connection process.

Installing the MS-Windows Version of Mosaic

This section shows you how to install the MS-Windows version of Mosaic. It includes a requirements checklist, instructions on how to obtain and install the software, and a discussion of potential problems.

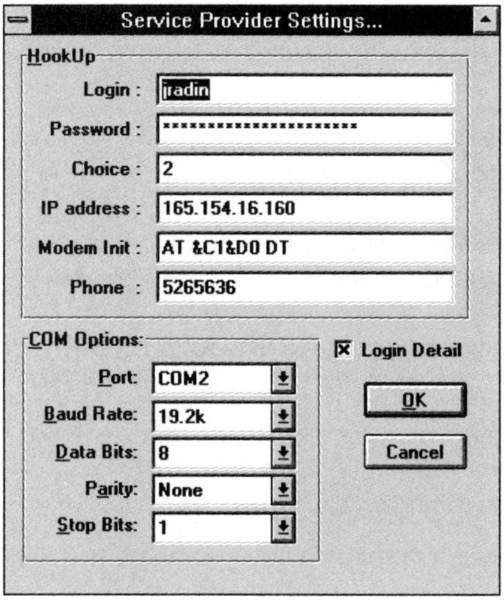

FIGURE B-9. *Service provider settings*

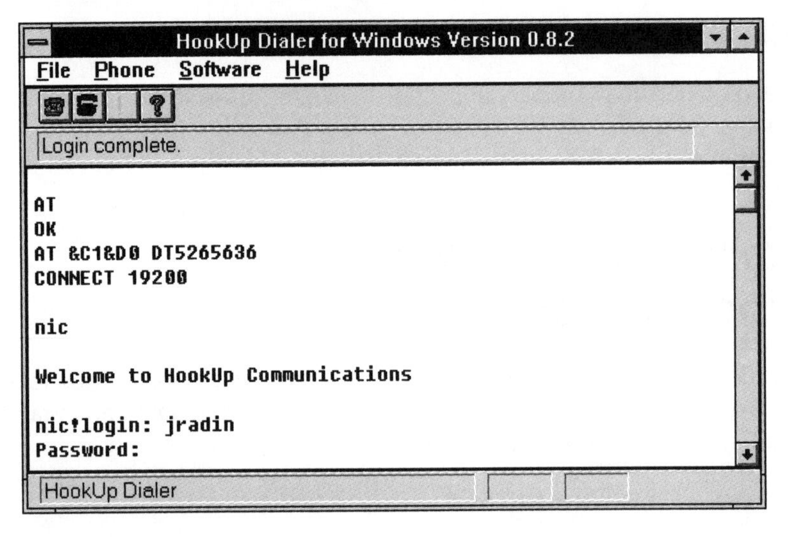

FIGURE B-10. *Sample login window*

Requirements Checklist

This step-by-step guide shows you how to install and configure NCSA Mosaic to run under Microsoft Windows. The following are required:

- The 32-bit software upgrade to Windows 3.1. (Windows NT already contains this software, so the upgrade is not necessary.)
- A direct Internet connection (TCP/IP connectivity) or access to the Internet provider over a PPP or SLIP connection by using a telephone line equipped with a modem.
- The **ftp** (file transfer protocol) utility to download files from the Internet.
- The **pkunzip** utility to unzip files stored in compressed format.
- An ASCII editor such as **edit** to customize the Mosaic configuration file (**mosaic.ini**), even if only to indicate your electronic mail address.

Obtaining the Software

The following procedure enables you to get a copy of the Mosaic software (Windows version) from the National Center for Supercomputing Applications (NCSA). It assumes a subdirectory name of **mosaic**, but you can use any subdirectory name.

First you need to make the Mosaic directory:

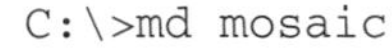

```
C:\>md mosaic
```

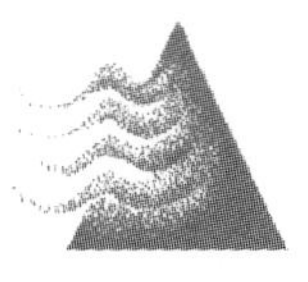

TIP
Create this subdirectory in the area normally used for installing applications on your system.

Then you can switch to the directory you just created using the following command:

```
C:\>cd mosaic
```

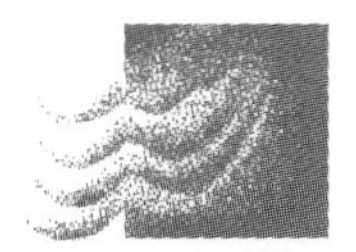

CAUTION
Pay strict attention to the case of characters that you enter in ftp commands; C and c are not equivalent.

Log on to NCSA's FTP server and download the NCSA Mosaic files.

```
c:\>ftp ftp.ncsa.uiuc.edu
```

At the login prompt, enter

```
login:ananymous
```

At the password prompt, enter your e-mail address, as in:

```
password:username@address
```

Wait for confirmation before continuing. Then enter

```
ftp>cd /Web/Mosaic/Windows
```

Enter the next command and read this valuable source of information before continuing.

```
ftp>get README.FIRST
```

Next you have a choice. Enter either

```
ftp>ls
```

or

```
ftp>dir
```

to list the available Windows files and directories.

To invoke binary mode for the file transfer, enter

```
ftp>bin
```

Then enter a command such as this one to retrieve the desired file:

```
ftp>get wmos20a7.zip
```

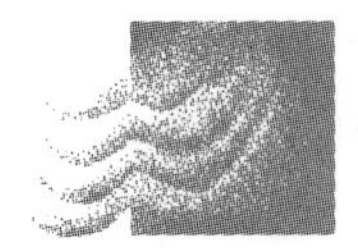

CAUTION

The file name **wmos20a7.zip** is subject to change. Its format is "wmos*version*.zip" where "wmos" is short for Windows Mosaic and "*version*" represents the current version number. For example, **wmos20a7.zip** is the file name for NCSA Windows Mosaic version 2.0, alpha release 7. An alpha release is the primary testing phase of software development for a given product.

The file **wmos20a7.zip** is a compressed archive containing the NCSA Mosaic executable and several documents, including an installation guide.

If you are running Windows NT, skip the steps that refer to the 32-bit software upgrade for Windows 3.1 mentioned earlier. Mosaic requires the win 32-bit software upgrade to Windows 3.1 **winsock.dll** (it must be WinSock 1.1-compliant). This software is found in the **win32s115.zip** file in the directory **/developer/devtools/win32sdk** of Microsoft's anonymous ftp server (**ftp.microsoft.com**) or in the NCSA anonymous ftp site (**ftp.ncsa.uiuc.edu**).

Connect to the server and issue a **get wins115.zip** command from the **ftp** prompt, for example

```
ftp>get win32s115.zip
```

The next command exits **ftp**, the file transfer protocol.

```
ftp>quit
```

You now have your files. Execute the **pkunzip** utility to retrieve and uncompress the archived file. Appendix C explains where to get this utility if you don't have it.

```
c:\mosaic>pkunzip wmos20a7.zip
```

If you are running Windows NT, you have finished the installation. Windows 3.1 users must issue the next command to receive the 32-bit software upgrade.

```
c:\mosaic>pkunzip wins32115.zip
```

Confirming the Files

Before going any further, check that all the following files exist in the **mosaic** directory:

- **update.txt** The current list of this version's enhancements and bug fixes in ASCII (industry standard) format.
- **install.txt** The NCSA Installation Guide for Mosaic for Windows in ASCII format.
- **install.wri** The NCSA Installation Guide for Mosaic for Windows in Microsoft Write (proprietary word processing) format.
- **mosaic.exe** The NCSA Mosaic executable file at the heart of the Mosaic system.
- **mosaic.ini** The Mosaic initialization and configuration file. This file is similar in function to the Microsoft Windows **win.ini** file. The **mosaic.ini** file was discussed in detail in Chapter 7.
- **readme.txt** Last minute or version introduction information.

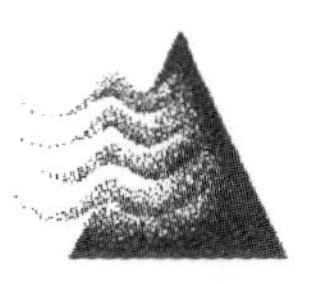

TIP
Read the **readme.txt** file before continuing. A small suggestion can save hours of headaches.

- **faq.txt** A list of Frequently Asked Questions (FAQs).

Once you are familiar with Mosaic, you will want to read **update.txt** as well. To see the entire enhancement list on line, access Mosaic, go to the NCSA Home Page, and then click on the Features List hyperlink.

Completing the Installation

The next step is to add a Mosaic icon to Microsoft Windows. Unlike some of the earlier ones, this procedure is short and sweet.

1. Access Microsoft Windows.
2. Choose New from the File menu.
3. Click on the Program Group button in the ensuing dialog box and click on the OK button to confirm.
4. Type **Mosaic** (or another description) in the ensuing dialog box and click on the OK button to confirm.

Potential Problems

If Mosaic hangs on execution or misbehaves in any other unexplained fashion, the most likely source of the problem is the WinSock DLL, software that is required to provide the TCP/IP networking under Windows. Check your Winsock documentation to see whether it is WinSock 1.1-compliant.

If you can execute NCSA Mosaic but cannot select the Windows Mosaic Home Page, try the following:

1. Select Open Local File from the File menu and try to open a file on your system.
2. Select Open URL from the File menu and try to open an HTTP file on remote Web server. The URL **http://www.ncsa.uiuc.edu/General/UIUC/UIUCIntro/Uofl_intro.html** is a good test file. Be careful when typing this name; make absolutely sure that you enter it correctly.
3. Select Open URL from the File menu and try to open an FTP file on a remote Web server. The URL file **//ftp.ncsa.uiuc.edu/PC/Mosaic/faq.txt** is a good test file.

Installing the X Window (Unix) Version of Mosaic

This section shows you how to install the X Window (Unix) version of Mosaic. It tells you how to obtain the software, and customize your installation.

Obtaining the Software

The following procedure enables you to obtain a copy of the Mosaic software (X Window version) from the National Center for Supercomputing Applications (NCSA).

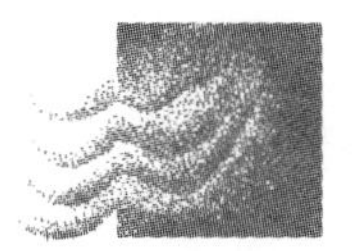

CAUTION
Pay strict attention to the case of characters that you enter in commands; C and c are not equivalent.

The following code accesses the NCSA FTP server to download the mosaic files:

```
$cd /usr/users/username
$mkdir bin
$cd /usr/users/username/bin
$ ftp ftp.ncsa.uiuc.edu
```

At the login prompt, enter

```
  login: anonymous
```

At the password prompt, enter your e-mail address, as in:

```
  password: username@address
```

Wait for confirmation before continuing. Then enter

```
ftp>bin
ftp>cd /bin
```

Next enter either

```
ftp>ls
```

or

```
ftp>dir
```

to list the available files, such as the following:

```
gunzip
gzip
zcat
uncompress
```

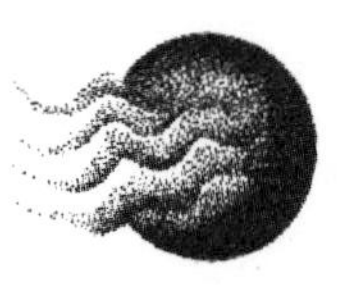

NOTE
Unix systems may already have some of the following files. Don't download them if you already have them.

```
ftp>get gunzip
ftp>get gzip
ftp>get uncompress
ftp>get zcat
```

Continue with the following entry:

```
ftp>cd /Web/Mosaic/Unix/binaries/2.4
```

Next enter either

```
ftp>ls
```

or

```
ftp>dir
```

to list the available files and directories.

The following binaries for major Unix versions such as DEC OSF/1 (Alpha) are available:

```
Mosaic-alpha.gz
Mosaic-dec.gz
```

```
Mosaic-hp6000.gz
Mosaic-ibm.gz
Mosaic-indy.gz
Mosaic-sgi.gz
Mosaic-solaris.gz
Mosaic-sun-lresolv.gz
Mosaic-sun.gz
Mosaic-alpha.gz
Mosaic-Readme-binaries
```

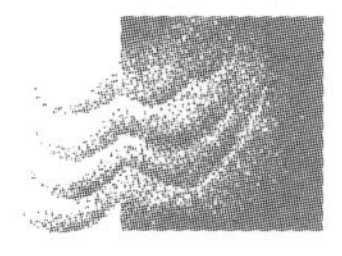

CAUTION
The **Mosaic-Readme-binaries** file lists the available Mosaic Unix version binary files. You may need to extract the source code in archived format before compiling it, using a generic MAKE file which requires specific parameters for your system. The moral of the story is that you should find out from your vendor if a compiled version of Mosaic is available.

Next retrieve the desired files such as **Mosaic-alpha.gz** for the DEC Alpha systems and exit **ftp**, the file transfer protocol.

```
ftp>bin
ftp>get Mosaic-Readme-binaries
ftp>get Mosaic-alpha.gz
ftp>quit
```

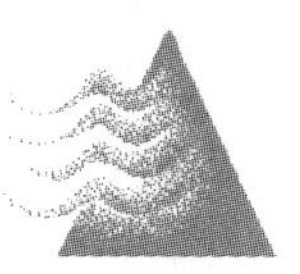

TIP
Under **/Web/Mosaic/Unix/Source/Mosaic_2.4/src** the file **Mosaic_2.4.tar.Z** will give the archived source files. Uncompress this file into the **Mosaic_2.4** directory.

Now uncompress the file (the example shows the version of Mosaic running on a DEC Alpha computer), change the file name to **xmosaic**, and change the file permissions as follows:

```
$ gunzip Mosaic-alpha.gz
$ mv Mosaic-alpha xmosaic
$ chmod +x xmosaic
$ chmod +w Mosaic
$ chown user-name xmosaic
$ chgrp users xmosaic
```

Customizing Your NCSA Mosaic

One size does not fit all. The **Mosaic** file contains settings for Mosaic. By editing this file you can change Mosaic's look and feel, and the search path for documents on the Internet. Users of all levels will modify this file to get an interface that they like.

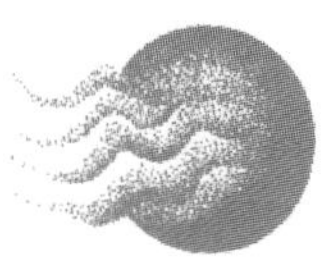

NOTE
You need to know about X Window to edit the **Mosaic** file extensively. If you are not familiar with X Window, consult *X Window Inside and Out*, by Levi Reiss and Joseph Radin (Osborne/McGraw-Hill, 1992).

Software installation is only the tip of the iceberg. What's important is testing it. Test Mosaic by entering the following:

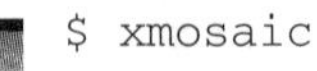

```
$ xmosaic
```

You should see a welcome screen.

If you are unable to start Mosaic, check with your network administrator that your computer has been properly configured.

Viewers and Players

One reason that so many people like Mosaic is its ability to use third-party software to view image files, movie files, and Postscript files, as well as to play audio files. In the MS-Windows version of Mosaic, the default **mosaic.ini** file includes the common MIME (Multipurpose Internet Mail Extensions) messaging standard that enables users to send multimedia documents across a TCP/IP network such as the Internet.

Default Settings

You should examine the Viewers and Suffixes sections of the **mosaic.ini** file to determine which viewers and players are defined by default.

Viewers Section

The first part of the Viewers section lists file types in the MIME format. The entries take the form TYPE*n*=, where *n* is an integer starting from 0. Here is a typical entry:

```
TYPE4="video/mpeg"
```

which basically means: use mpeg files for video.

The second part of the Viewers section specifies the full path to the viewer for each listed file type, as in:

```
video/mpeg="c:\windows\apps\mpegplay %ls"
```

The *%ls* parameter is replaced with a specific file name on the command name.

Suffixes Section

The Suffixes section lists file name suffixes, or extensions, identifying the file types for files. There is no limit on the number of file name extensions for a given file type. The extensions are separated by commas. The last extension in the list is used when you write a file to the local hard drive.

Here is the entry that corresponds to the selected entries in the Viewers section:

```
video/mpeg=.mpeg,.mpe,.mpg
```

This entry denotes that files with the extension .mpeg, .mpe, or .mpg are MPEG images—in other words, movies.

Finding Viewers

Mosaic for Microsoft Windows uses external viewers and players to display certain file types, such as JPEG images or MPEG movies. You can download these viewers and players from the NCSA FTP server in the directory **/WebMosaic/Windows /viewers**.

Information about viewers tested to run with Mosaic is available via the following URL:

http://www.ncsa.uiuc.edu/SDG/Software/WinMosaic/viewers.html

Figure B-11 shows a list of tested viewers. To download a viewer—Mpegplay, for example—just hold down SHIFT and click on the link. These viewers are also available on the NCSA anonymous ftp server in the **/Web/Mosaic/Windows/ viewers** directory. The viewers are in .zip format so you have to unzip them, as described earlier in this appendix, before you can use them.

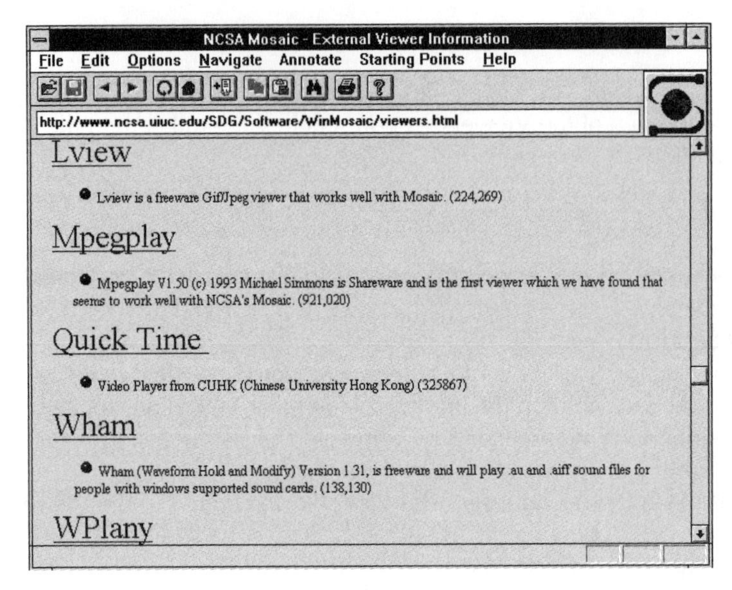

FIGURE B-11. *Information about viewers*

Installing MS-Windows Viewers

The example in this section defines a new viewer that reads .wav files. To do so, use any ASCII editor such as Window's Notepad (in the Accessories program group) to edit the Viewers section and the Suffixes section of the **mosaic.ini** file. After making these changes, reload Mosaic so that the changes take effect.

Modifying the Viewers Section

Add the following entry to the first part of the Viewers section:

```
TYPEn="video/mpeg"
```

where *n* is the integer following the last existing TYPE entry in the Viewers section.

Then add an entry such as

```
video/mpeg="c:\path_to_your_viewer\XXXX %ls"
```

to the second part of the Viewers section. Add this entry after the TYPE*n* entry, but before the telnet= entry.

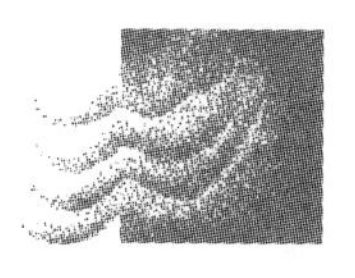

CAUTION
The viewer program must be in your **windows** directory or in a directory defined in the path statement of your **autoexec.bat** file—for example, "video/mpeg=c:\winapps\mpeg\mpegplay.exe %ls."

The tested viewers include Lview (an image viewer stored in **lview31.zip**), Mpegplay (an image viewer stored in **mpegw32e.zip**), QuickTime (a video player stored in **qtw11.zip**), Wham (a sound player stored in **wplny09b.zip**), Wplany (a sound player stored in **wplny09b.zip**), and GhostScriptv2.6 (a Postscript viewer stored in **gs261exe.zip** and **gsview10.zip**).

Modifying the Suffixes Section

After you finish changing the Viewers section, list the extensions associated with this new type in the Suffixes section. For example, you could enter

```
audio/wav=.wa,.wav
```

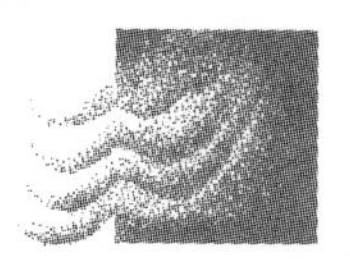

CAUTION
Don't forget to restart Mosaic so that the changes will take effect and the viewers will become available.

Installing Unix Viewers

This section illustrates how to install viewers under Unix. The process tends to be more complicated than installing viewers under the MS-Windows version of Mosaic.

First make the appropriate directory and get into that directory:

```
$mkdir /usr/users/username/viewers
$cd /usr/users/username/viewers
```

Then connect to the NCSA ftp server and identify yourself:

```
$ftp ftp.ncsa.uiuc.edu
login:anonymous
password:username@address
```

Next change to the directory that will contain the viewers and examine the list of available viewers.

```
ftp>cd /Web/Mosaic/Unix/viewers
ftp>dir
   ghostview-1.4.1.tar.Z
   ghostscript-2.6.tar.Z
   ghostscript-fonts-2.6.tar.Z
   mpeg-play-2.0.tar.Z
   xv-3.00.tar.Z
   xdvi.tar.Z
ftp>bin
```

Then get, for example, the widely employed xv viewer that helped create the Unix screen dumps in this text:

```
ftp>get xv-3.00.tar.Z
```

Now switch directories and obtain compressed copies of a GUI used with the external sound player showaudio or the movie player mpeg_play:

```
ftp>cd xplaygizmo
ftp>dir
   xplaygizmo-sgi.Z
   xplaygizmo-sun.Z
   xplaygizmo.aplha.Z
   xplaygizmo.hp.Z
   xplaygizmo.ibm.Z
   xplaygizmo.readme
   xplaygizmo.release-notice
```

Next get the appropriate xplaygizmo-related files and exit ftp:

```
ftp>get xplaygizmo.readme
ftp>get xplaygizmo.release-notice
ftp>get xplaygizmo.alpha.Z
ftp>quit
```

Now you can uncompress your files and extract xv using utilities commonly available in Unix :

```
$uncompress xplaygizmo.alpha.Z
$uncompress xv-3.00.tar.Z
$tar xvf xv-3.00.tar
```

Then copy the appropriate xplaygizmo file into the user directory and change permissions:

```
$cp xplaygizmo.alpha /usr/users/username/xplaygizmo
$chmod +x /usr/users/username/xplaygizmo
```

At this point, repeat this process for the xv file.

```
$cd xv-3.00
$cp xv /usr/users/username
$chmod +x /usr/users/username/xv
```

In the directory **/usr/users/username** create a **.mailcap** file to link MIME types to external viewers with the following lines:

```
audio/*; xplaygizmo showaudio %s
video/mpeg; xplaygizmo mpeg_play %s
```

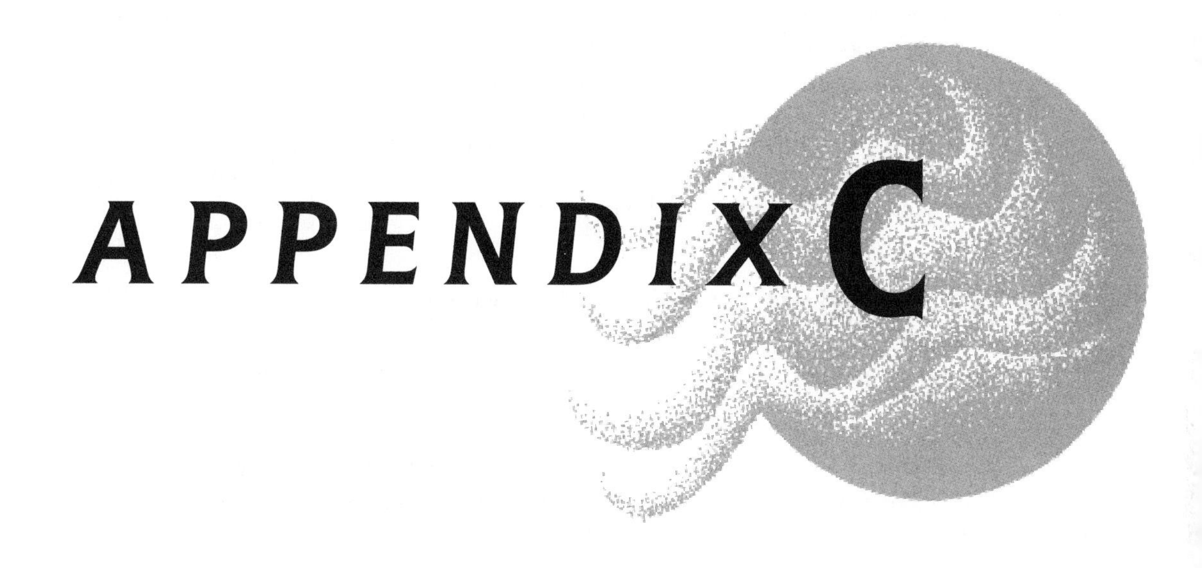

The Server

This appendix is for users who want a technical discussion regarding the installation and configuration of the server. In the client-server environment, the server is needed to centralize and provide (serve) information to clients. In essence, all the documentation is on the servers; the clients go out and get it. As you know, Mosaic is a client application. It browses a server such as the World Wide Web server. While several server versions are available—such as NCSA and CERN—this text discusses only the NCSA Unix and MS-Windows versions. You will see how similar they are.

To run the server, you need to install a process (program) called httpd on the computer. The two server versions share many features. They support scripts that generate documents, they support queries, and they accept user inputs. The security function limits directory access. You can use this feature to render your server more secure.

You don't need to make any major changes to the server configuration files. As shown later in this appendix, you simply install these files into the appropriate directory as per the installation instructions. NCSA httpd will serve documents to World Wide Web browsers. Directory indexes let users know contents of your directories. An HTML listing of your directories can be stored in a document or generated by the server upon demand. When generating the listing, the server can add administrator-defined descriptive strings to directory entry items, and mark them with icons depending on the file type.

MS-Windows Server Features

Version 1.3 is a release of NCSA httpd for Microsoft Windows, written by Bob Denny. The code is based on the Unix NCSA httpd Version, written by Rob McCool.

NCSA httpd for Windows is fully compliant with the WinSock 1.1 specification and so works with most Windows-based TCP/IP packages, such as Chameleon and Trumpet.

Installing the Server

The server installation process is discussed separately for the two versions. After downloading the software, you must configure the three server files: the main configuration file (httpd), the configuration file (access), and the resource configuration file (srm).

We can divide the installation process of NCSA httpd for Windows into several basic steps: downloading the software, setting up the software, configuring the server, and starting the server.

Downloading the MS-Windows Version

The first thing you should do is edit your **autoexec.bat** file using the text editor:

```
c:\> edit autoexec.bat
```

Add the following line to set the time zone variable:

```
c:\>SET TZ=EST4EDT
```

where 4 denotes 4 hours west of GMT. If you don't use daylight savings time, omit the second set of three letters.

Reboot the computer, create the httpd directory, and switch to that directory:

```
c:\>md httpd
c:\>cd httpd
c:\httpd>ftp ftp.alisa.com
```

To get the full version of the server, do the following:

```
login: anonymous
password: username@address
ftp>cd /pub/win-httpd
ftp>bin
ftp>get whttpd13.zip
ftp>quit
```

The following line extracts the server with the *-D* option.

```
c:\httpd>pkunzip -D whttpd13.zip
```

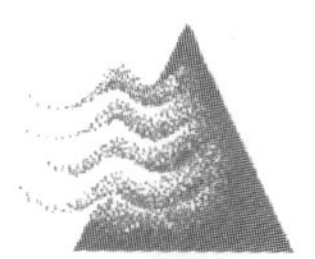

TIP

The -D option preserves directories. It is very important.

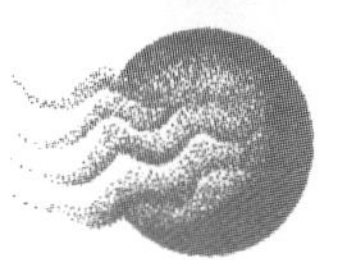

NOTE

Use **pkunzip** version 2.04g or later to uncompress the software. To find this software, log in to **ftp.ncsa.uiuc.edu** as an anonymous user and get this version from the **/PC/Windows/Contrib** directory. You need to get the **pkzip204g.exe** self-extracting program that generates the desired pkunzip version.

An alternative site for downloading the MS-Windows version is **ftp.ncsa.uiuc.edu**. You will find the full version of this server in the directory **/Web/httpd/Unix/ncsa_httpd/contrib/winhttpd.**

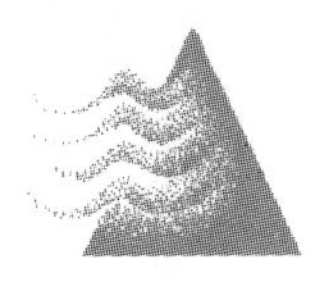

TIP
Some Internet connection packages such as Chameleon include an ftp Graphical User Interface for downloading files, as shown in Figure C-1. It is easier to use this GUI instead of typing commands.

Downloading the Unix Version

To download the Unix version, execute the following commands:

```
$cd /usr/local/etc
$ftp ftp.ncsa.uiuc.edu
login:anonymous
password:username@address
ftp>cd /Web/httpd/Unix/ncsa_httpd/current
ftp>bin
ftp>dir
```

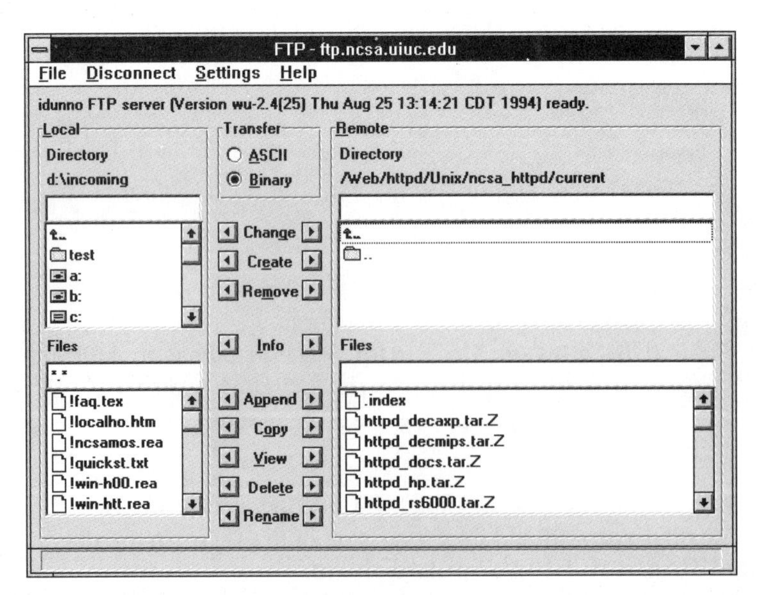

FIGURE C-1. *The Chameleon ftp interface for downloading files*

Then, depending on your system, enter one of the following:

```
# The following line is for Silicon Graphics
httpd_sgi.tar.Z
# The following line is for Sun Microsystems
httpd_sun.tar.Z
# The following line is for DEC MIPS
httpd_decmips.tar.Z
# The following line is for DEC Alpha
httpd_decaxp.tar.Z
# The following line is for IBM RS6000
httpd_rs6000.tar.Z
# The following line is for Hewlett-Packard
httpd_hp.tar.Z
```

The following example is for the DEC Alpha system:

```
ftp>get httpd_decaxp.tar.Z
```

In all cases you will need the documentation, which you can access with a command such as the following:

```
ftp>get httpd_docs.tar.Z
ftp>quit
```

Uncompress and extract the files for the DEC Alpha system:

```
$uncompress httpd_decaxp.tar.Z
$tar xvf httpd_decaxp.tar
```

The preceding set of commands creates the **httpd** directory, whose complete path is **/usr/local/etc/httpd**. This directory includes all software associated with the server.

```
$ uncompress httpd_docs.tar.Z
$ tar xvf httpd_docs.tar
```

The preceding set of commands creates the **docs** directory, whose complete path is **/usr/local/etc/httpd/docs**. This directory includes all documentation associated with the server.

If you don't find the compressed file for your specific computer on the ftp server, you should download the **httpd_source.tar.Z** file, extract it, and compile to create the server program. To do so, use code such as the following:

```
ftp>get httpd_source.tar.Z
ftp>quit
$ mkdir /usr/local/etc/httpd_src
$ cd /usr/local/etc/httpd_src
$ uncompress httpd_src.tar.Z
$ tar xvf httpd_source.tar
```

The preceding set of commands creates the **httpd_src** directory, whose complete path is **/usr/local/etc/httpd_src**. This directory includes all software associated with the server.

Configuring the Server

For both versions, the three server files are the main server configuration file, the access configuration file, and the server resource configuration file.

UNIX
Configuring Unix server files requires the superuser (root) account.

The following rules apply to all httpd configuration files:

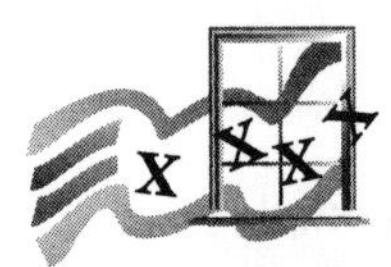

UNIX
The configuration files are case sensitive.

WIN
The configuration files are not case sensitive.

- Comment lines begin with the pound sign (#).

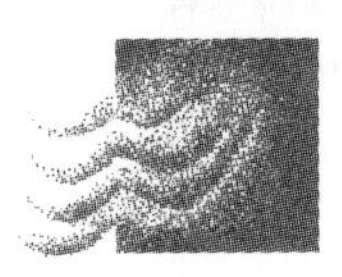

CAUTION
The pound sign (#) must be the first character on the line. Comments must be on a line by themselves.

- Use one directive per line. A directive is a key word that httpd recognizes, followed by one space and data specific to the directive. For example, one line might read

  ```
  Port 80
  ```

- Extra white space is ignored. To embed a space in data without separating it from any subsequent arguments, use a \ (backslash) character before the space.

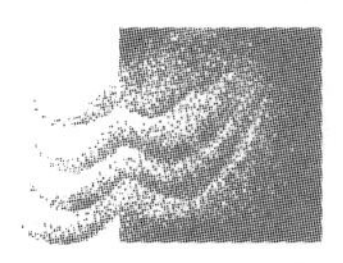

CAUTION
Don't change the order of the file directives. If you don't understand the instructions, consult the online documentation.

The Main Server Configuration File

The main server configuration file is quite similar for the two server versions. The configuration file for the Unix version is

```
httpd.conf
```

and its location is

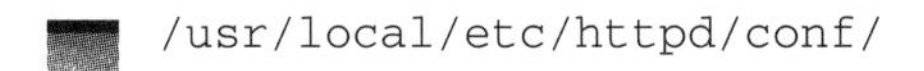

```
/usr/local/etc/httpd/conf/
```

while the configuration file for the Windows version is

```
httpd.cnf
```

and its location is

```
c:\httpd\conf\
```

UNIX
The order of directives is as follows: Server Type, Port, User/Group, ServerAdmin, ServerRoot, ErrorLog, TransferLog, PidFile, and ServerName.

WIN
The order of directives is as follows: Server Root, Port, Timeout, ServerAdmin, ErrorLog, TransferLog, PidFile, and ServerName.

ServerRoot is the directory the server's config, error, and log files are kept in. This should be specified in the startup command line.

The Server Root directive for the two systems follows:

WIN
ServerRoot c:/httpd/

UNIX
ServerRoot /usr/local/etc/httpd

Port is the port the stand-alone listens to. Use

```
Port 80
```

UNIX
Ports whose number is less than 1023 as defined in the ***/etc/services*** file require running the httpd server program as a privileged user initially.

Timeout is the timeout applied to all network operations. If you are on a slow network, or are using a SLIP or PPP connection, try increasing this setting to 60 seconds. In our case we'll use

```
TimeOut=30
```

When users have problems with the server they should send e-mail to the address specified by the ServerAdmin directive.

```
ServerAdmin jradin@hookup.net
```

ErrorLog is the location of the error log file.

```
ErrorLog logs/error.log
```

TransferLog is the location of the transfer log file.

```
TransferLog logs/transfer.log
```

ServerName allows you to set a host name that is sent back to clients for your server. There is no default ServerName. Here's an example:

```
ServerName jradin.ott.hookup.net
```

UNIX
ServerType is either inetd (a process that monitors all network operations) or standalone. If you are running from inetd, go to "ServerAdmin." In this case, we use

```
ServerType standalone
```

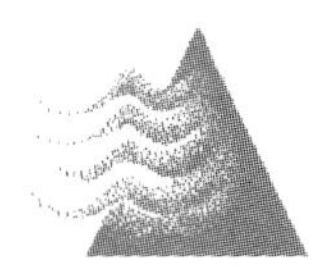

TIP
A standalone server increases performance because it doesn't have overhead.

UNIX
User/Group is the name (or #number) of the user/group allowed to run httpd. Our example uses

```
User joe
Group users
```

UNIX
PidFile is the file the server should log its pid to.

```
PidFile logs/httpd.pid
```

The Access Configuration File

The access configuration file is quite similar in the two server versions. It establishes unrestricted access to the server's document tree.

The configuration file for the Unix version is

```
/usr/local/etc/httpd/conf/access.conf
```

while the configuration file for the Windows version is

```
c:\httpd\conf\access.cnf
```

Change the following to your setting of the ServerRoot as defined in the main server configuration file:

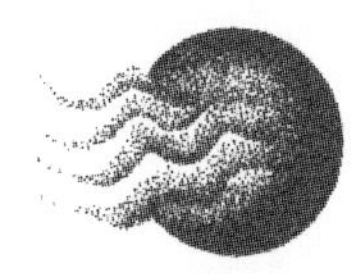

NOTE
The brackets are part of the HTML syntax as discussed in Chapters 5 and 6.

WIN
<Directory c:/httpd>
Options Indexes
</Directory>

UNIX
<Directory /usr/local/etc/httpd/>
Options Indexes FollowSymLinks
</Directory>

Change the following to your setting of the DocumentRoot as defined in the server resource configuration file discussed below:

WIN
<Directory c:/httpd/htdocs>
This may also be "None," "All," or "Indexes."

UNIX
<Directory /usr/local/etc/httpd/htdocs>
This may also be "None," "All," or any combination of "Indexes," "Includes," or "FollowSymLinks."

WIN
Options Indexes are directives that control which options the **haccess.ctl** files in the document directories can override. Options Indexes can also be "None," or any combination of "Options," "FileInfo," "AuthConfig," and "Limit."

UNIX
Options Indexes FollowSymLinks are directives that control which options the **.htaccess** files in the document directories can override.

Options Indexes can also be "None," "All," or any combination of "Indexes," "Includes," or "FollowSymLinks."

AllowOverride All controls who can retrieve files from this server. This example allows access by all:

```
<Limit GET>
order allow,deny
allow from all
</Limit>
</Directory>
```

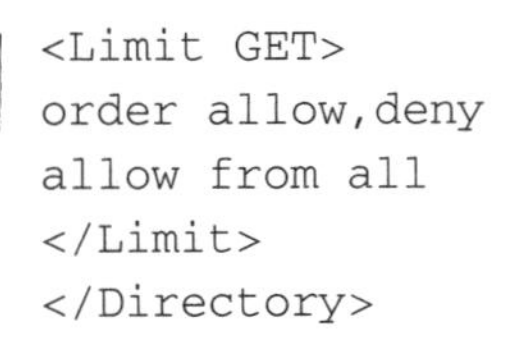

NOTE
You may place any additional access document directories after the </Directory> statement.

The Server Resource Configuration File

The server resource configuration file is quite similar for the two server versions. This file controls the document layout and name specs that your server makes visible to users.

The configuration file for the Unix version is

```
/usr/local/etc/httpd/conf/srm.conf
```

while the configuration file for the Windows version is

```
c:\httpd\conf\srm.cnf
```

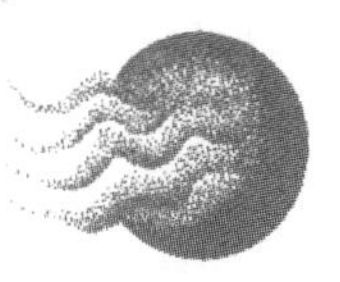

NOTE
Path defaults are set according to the server's installation directory (ServerRoot). Paths may be given in Unix or DOS format (using either / or \).

DocumentRoot is the directory that serves your documents. By default, all requests access this directory, but symbolic links and aliases may point to other locations.

WIN
DocumentRoot c:/httpd/htdocs

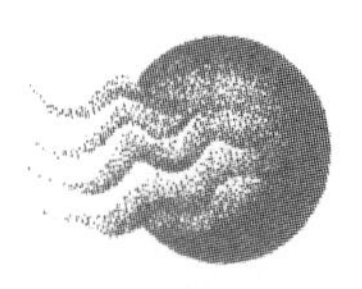

UNIX
DocumentRoot /usr/local/etc/httpd/htdocs

UserDir denotes the directory appended onto a user's home directory if a request is received. User-relative URLs are not supported in this version.

UNIX
UserDir public_html

DirectoryIndex is the name of the file to use as a prewritten HTML directory index.

WIN
DirectoryIndex index.htm

AccessFileName denotes the file in each directory containing access control information.

WIN
AccessFileName haccess.ctl

UNIX
AccessFileName .htaccess

Redirect allows you to tell clients where to find documents that no longer exist on your server.

```
Redirect fakename url
```

Aliases are short names that replace longer ones. Up to 20 aliases are allowed. One useful alias specifies the path to the icons used for the server-generated directory indexes. The paths given here in the AddIcon statements are relative.

WIN
Alias /icons/ c:/httpd/icons/

UNIX
Alias /icons/ /usr/local/etc/httpd/icons/

ScriptAlias controls which directories contain server scripts.

WIN
ScriptAlias /cgi-bin/ c:/httpd/cgi-bin/

UNIX
ScriptAlias /cgi-bin/usr/local/etc/httpd/cgi-bin/

MIME Content Type Control includes three directives: DefaultType, AddType, and ReadmeName. DefaultType denotes the default MIME type for documents whose type the server cannot find from the file name extensions.

```
DefaultType text/plain
```

AddType allows you to set selected files to given types; for example the following line sets files whose extension is ext1 to the file type subtype.

```
AddType type/subtype ext1
```

ReadmeName denotes the default **README** file accessed by the server. The server first looks for **name.html**, includes it if it's found, and then looks for name and includes it as a plain text (ASCII) file if it's found.

```
ReadmeName README.
```

FancyIndexing denotes standard or fancy directory indexing.

FancyIndexing on AddIcon specifies the icon for different files or file name extensions.

TIP
In preparation for the upcoming Windows version, include explicit three-character truncations for four-character endings, instead of relying on DOS to truncate for you.

The following directives tell the server which icons to show for different files or file name extensions.

WIN
AddIcon /icons/text.xbm .html .htm .txt .ini

UNIX
AddIcon /icons/text.xbm .html .txt

WIN
AddIcon /icons/image.xbm .gif .jpg .jpe .jpeg .xbm .tiff .tif .pic .pict

UNIX
AddIcon /icons/image.xbm .gif .jpg .xbm .tiff

WIN
AddIcon /icons/sound.xbm .au .wav .snd

UNIX
AddIcon /icons/sound.xbm .au

WIN
AddIcon /icons/movie.xbm .mpg .mpe .mpeg

WIN
AddIcon /icons/movie.xbm .mpg

WIN
AddIcon /icons/binary.xbm .bin .exe .bat .dll

UNIX
AddIcon /icons/binary.xbm .bin

```
AddIcon /icons/back.xbm
AddIcon /icons/menu.xbm
```

DefaultIcon denotes the default file icon. For example:

```
DefaultIcon /icons/unknown.xbm
```

AddDescription provides a short description of a file in server-generated indexes. For example:

```
AddDescription "description" filename
```

IndexIgnore denotes file names that directory indexing should ignore.

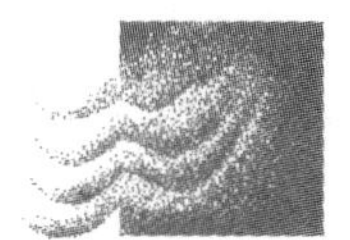

CAUTION
Set an ignore for your access control file(s):
IndexIgnore / ! ~

Starting the Server

The server displays a "splash screen" at startup. When the initialization is complete, the server shrinks into an icon. If a user connects and requests a document from the minimized server, the icon title changes to "httpd active" and the glowing balls travel down the "S".

Figure C-2 provides suggestions for starting up the Unix server. Start the server and perform initial tests.

The MS-Windows server startup check list follows:

1. Create a server icon by dragging **httpd.exe** from the File Manager window to a Program Manager group window. The icon name is MSHTTPD13.
2. Make sure that Winsock is running. The server uses DNS (Domain Name Service) to identify itself and the client host names for logging purposes.
3. Double-click the httpd icon. If DNS is unavailable, initialization may take 30 seconds to complete, depending on the TCP/IP package.

After the server is minimized to an icon, it is ready for use. Try to open your Home Page. If you cannot, try to get localhost to work. Without it, most of the script and forms examples will fail.

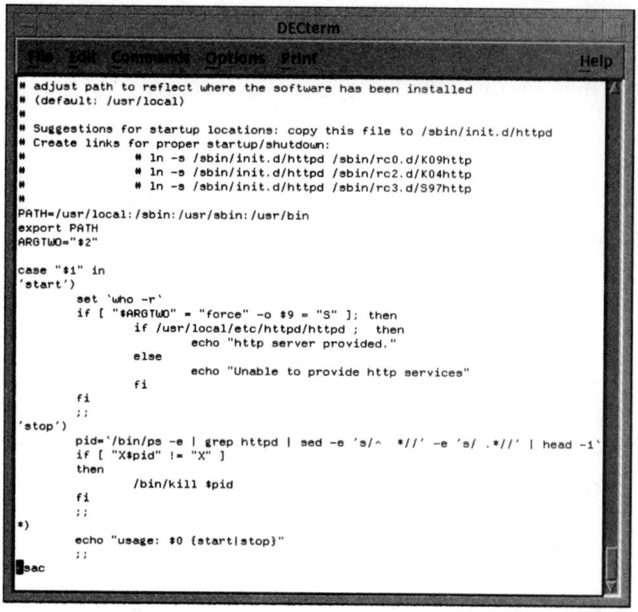

```
DECterm
File  Edit  Commands  Options  Print                                Help
# adjust path to reflect where the software has been installed
# (default: /usr/local)
#
# Suggestions for startup locations: copy this file to /sbin/init.d/httpd
# Create links for proper startup/shutdown:
#               # ln -s /sbin/init.d/httpd /sbin/rc0.d/K09http
#               # ln -s /sbin/init.d/httpd /sbin/rc2.d/K04http
#               # ln -s /sbin/init.d/httpd /sbin/rc3.d/S97http
#
PATH=/usr/local:/sbin:/usr/sbin:/usr/bin
export PATH
ARGTWO="$2"

case "$1" in
'start')
        set `who -r`
        if [ "$ARGTWO" = "force" -o $9 = "S" ]; then
                if /usr/local/etc/httpd/httpd ;  then
                        echo "http server provided."
                else
                        echo "Unable to provide http services"
                fi
        fi
        ;;
'stop')
        pid=`/bin/ps -e | grep httpd | sed -e 's/^  *//' -e 's/ .*//' | head -1`
        if [ "X$pid" != "X" ]
        then
                /bin/kill $pid
        fi
        ;;
*)
        echo "usage: $0 {start|stop}"
        ;;
esac
```

FIGURE C-2. *Unix server startup script*

The following command-line flags are available and may be set in the command line via the Program Manager¦File¦ Properties menu selection. A Program Item Properties dialog box is displayed on the screen. Append the appropriate flags to the command in the Command Line text box.

- The -d directory specifies a server root other than **c:\httpd**. This indicates where httpd will look for its configuration files.
- The -f file specifies the startup server configuration file.

Figure C-3 shows Joseph's Home Page, which indicates the link to the MS-Windows server, which is up and running. Click on the server hyperlink to generate Figure C-4.

Shutting Down the Server

Don't be in a rush to shut down the server. Users may be reading a document consisting of multiple linked pages and inline images or sounds, each of which is

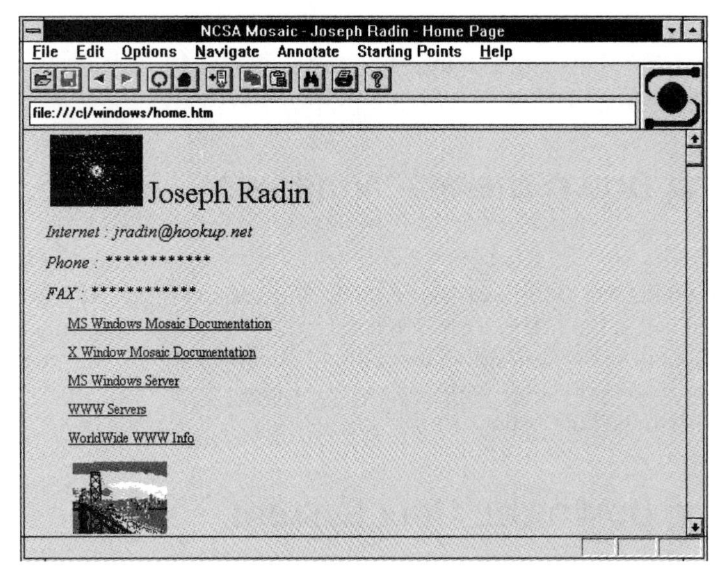

FIGURE C-3. *A Personal Home Page*

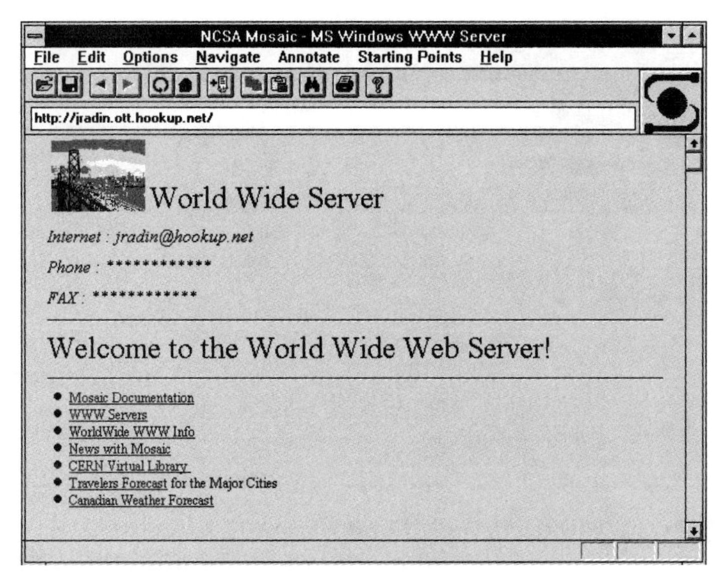

FIGURE C-4. *A MS-Windows Server Home Page*

read with a separate transaction. The server will not shut down while a transaction is in progress, but waits for the transaction to be completed. Some browsers behave unpredictably if the server is shut down unexpectedly.

Shutting Down the MS-Windows Server

To shut down the MS-Windows server, click twice on the httpdidle icon (outside the program group box). This icon is maximized. Click on the Control menu item, and select Exit from the pull-down menu that appears as shown in Figure C-5. The server beeps to acknowledge your request, waits until no transactions are in progress, and then shuts down.

Shutting Down the Unix System Server

To shut down the Unix system server, issue the following command:

```
$ ps -ef ¦ grep httpd
```

FIGURE C-5. *Shutting down an MS-Windows server*

As a result you will get the process ID, for example 1028. Then you kill this process. For instance, you might enter

```
$ kill 1028
```

Index

A

About HoTMetaL, 136-137
About HoTMetaL Pro, 137
Access configuration file, 241-243
Access guidelines (Internet), 16
Address elements, HTML, 118
Addresses, Internet, 13-14. *See also* URLs
Advanced Research Projects Agency, 3
Aliases, 244
ALIGN element (in HTML), 119
Alt.bbs, news on, 110
Alternative Internet newsgroups, 109-110
Anchor elements (in HTML), 117
Anchor variables in mosaic.ini file, 150-152
Annotate menu (Mosaic for Windows), 31-32, 177
Annotate menu (Mosaic for X Window), 49-51, 183-184
Annotate window, 32, 50, 157, 178
Annotations
    hyperlinks to, 178
    personal, 32, 178-179
    personal vs. group, 156
Application window, size and cursor, 152
Archie, 7-8
ARPAnet, 3
ASCII text files, e-mail as, 6
Autoload Home Page variable in mosaic.ini file, 146

B

Back icon, 64
Background (window), setting, 149
Bay Area Rapid Transit System, 102-103
Bay Area Restaurant Guide, 101-102
Billing, 15
Binary files, e-mail, 6
Blockquote tags (in HTML), 118

BODY elements (in HTML), 115
Boldfacing tags (in HTML), 119
Bookmarks (Gopher), 8, 99
Bug list, 149
Bug list address, setting, 147-148
Bug List variable in mosaic.ini file, 147-148
Bulleted list tags (in HTML), 118
Bullets, setting type of, 150
Business applications for the Internet, 10-11
Buttons, 184-188
Buying a home, 4-5

C

Cables, 13
California WWW Servers document, 101
CERN Web, 8-10
Chameleon, 12, 56, 205-214
- accessing the FTP program, 209, 213-214
- accessing a remote system, 213
- Final Settings dialog box, 212
- ftp interface for downloading files, 236
- hardware configuration, 207
- installation, 208
- Login Settings dialog box, 211
- mail and news settings, 208
- modem settings, 207
- Modem Settings dialog box, 211
- network connection process, 212
- Open Configuration File window, 209-210
- Port Settings dialog box, 210
- using dial-up connections, 208-209
- Windows program group, 209

Character dialog box (HoTMetaL), 135
Character highlighting in HTML, 119
Character strings, copying or finding, 26, 44
Client application, Mosaic as, 233
Client browser, Mosaic as, 56
Clients, 17
Client-server architecture, 17
CMU School of Computer Science Home Page, 69-70
Coaxial cables, 13
Comet watch (Shoemaker-Levy 9), 60-64
Comments
- in HTML, 119
- in mosaic.ini file, 145
- types of, 31, 49, 51

Commercial WWW servers, 58
Communications protocols, 12-13
Communications software, 12, 56, 155, 205
CompuServe gif files, viewing, 77-78
Configurations, temporary vs. permanent, 170
Configuring Mosaic, 143-167
Configuring Mosaic for Windows, 143-163
Configuring Mosaic for X Window (Unix), 163-166
Configuring the server, 237-247
Configuring Trumpet Winsock, 215-216
Connecting to the Internet, 22-23, 40, 205-206
Connecting to Mosaic, 56-57
Copying a character string, 26
Corporate Home Pages, 91, 93-94 (*see also* Home Pages)
- creating, 138-140
- sample, 92-94

Cursor, application window, 152
Customizing Mosaic, 143-167

D

Data encryption, 16
Debugging info menu (Mosaic for Windows), 30
DEC (Digital Equipment Corporation)
- customer services, 94

Home Page, 92-93
servers, 4, 6
Default Title variable in mosaic.ini file, 152, 156
Direct connection to the Internet, 205-206
Directory variable in mosaic.ini file, 156
Display Inline Images variable in mosaic.ini file, 148-149
Document Source dialog box (Mosaic for X Window), 45
Document titles
removing from the screen, 74-75
setting display of, 151
Documentation, online, 162-164
DocumentRoot directory, 243
Documents (WWW)
adding to a Hotlist, 59
identifying and creating, 82
previewing, 129, 172
saving, 26
saving to a local disk, 173
Domain, 13
Domain names, 13-14
DOS, TCP/IP for, 13
Downloading, explained, 11

E

Edit menu (HoTMetaL), 130-131
Edit menu (Mosaic for Windows), 26-27, 171-172
Educational Home Pages, 91-92, 94-95
Electronic mail. *See* E-mail
Elements, HTML, 114-120
E-mail, 6, 11
E-mail address, setting, 146
E-mail messages, 6
E-mail text files, 6
E-mail variable in mosaic.ini file, 146
Embedded images in HTML documents, 119-120
English Server, 69, 71
English Server Home Page, 70
Entities of HTML elements, 114
Errors, troubleshooting, 166-167
Etiquette, online, 15
European Center for Nuclear Research (CERN), 9
Extended FTP variable in mosaic.ini file, 151

F

Failed DNS Lookup error, 166
Fancy Rules variable in mosaic.ini file, 150
FAQ (Frequently Asked Questions) Page, 147-148
FAQ Page variable in mosaic.ini file, 147
Feature Page, 148-150
Feature Page variable in mosaic.ini file, 148
Fiber-optic cables, 13
File menu (HoTMetaL), 126-130
File menu (Mosaic for Windows), 25-26, 170-171
File menu (Mosaic for X Window), 42-46, 180-182
File transfer protocols. *See* Ftps
File transfers, 6
Final Settings dialog box (Chameleon), 212
Find dialog box (Mosaic for Windows), 27
Find In Document dialog box (Mosaic for X Window), 44
Find & Replace dialog box (HoTMetaL), 132
Find and Replace option (HoTMetaL), 131
Finding a character string, 26
Firewall (computer), 16
Font selection, 29
Font window (Mosaic for Windows), 175
FTP archives, 6

Ftp command, 6
FTP program, accessing with Chameleon, 209, 213-214
Ftp (file transfer protocol) URL, 83
Ftps (file transfer protocols), 6
Ftp.ncsa.uiuc.edu, 22, 40, 153
Full-text search, 105

G

Geographical top-level domain names, 14
Gif files, viewing, 77-78
GolfData On-Line Home Page, 66
Golfing, 66-67
Gopher, 8, 71, 99
Gopher menu, 9
Gopher servers, 99, 197
Gopher Servers menu (Mosaic for Windows), 34
Gopher URL, general form for, 83
Graphic images, setting display of, 148-149
Grey Background variable in mosaic.ini file, 149
Group Annotation Server in mosaic.ini file, 156-157
Group Annotations variable in mosaic.ini file, 156
Group comments, 31, 51

H

HEAD elements (in HTML), 115-116
Help menu (HoTMetaL), 136
Help menu (Mosaic for Windows), 33, 35-36, 180
Help menu (Mosaic for X Window), 51, 184
Help Page address, setting, 147
Help Page variable in mosaic.ini file, 147
History feature, 97-98
History of the Internet, 3
History option (Mosaic for Windows), 32
Home buying, 4-5
Home Page address, setting, 146-147
Home Page variable in mosaic.ini file, 146-147
Home Pages (*see also* Corporate Home Pages; Personal Home Pages)
  internal view of, 27
  Mosaic for X Window, 41
  NCSA Mosaic, 23-24
  NCSA URL for, 82
  two-directional, 95
  types of, 84-96
  virtual, 92
Honolulu Community College Home Page, 73-74, 76
HookUp Communications software, 56, 217-218
Hotlist View dialog box (Mosaic for X Window), 48
Hotlists, 59, 97
HoTMetaL
  About HoTMetaL, 136-137
  Character dialog box, 135
  configuring, 125-126
  creating HTML documents with, 120-138
  downloading the software, 121-122
  Edit menu, 130-131
  File menu, 126-130
  find and replace, 131-132
  Help menu, 136
  installing, 122-126
  Markup menu, 134-136
  menus, 126-138
  Open Ascii Styles dialog box, 135
  Open dialog box, 127
  Open template dialog box, 128
  Preview option, 129
  running, 125
  View menu, 132-134

Window menu, 138
HoTMetaL Pro, 121, 137
HP online help, 11-12
HT Access error, 167
HTML elements, 114-120
ALIGN, 119
BODY, 115
HEAD, 115-116
IMG, 119-120
TITLE, 116
HTML (Hypertext Markup Language), 84, 113-141
address elements, 118
anchor elements, 117
blockquote tags, 118
character highlighting, 119
comments tags, 119
creating documents with HoTMetaL, 120-138
downloading the manuals, 120
embedded images in documents, 119-120
explained, 82
getting information on, 120
heading levels, 116-117
lists tags, 118
paragraph marks, 117-118
preformatted text tags, 118
source code for NCSA Mosaic Home Page, 114
structure, 115
tags, 114-115
text formatting elements, 117-119
Httpd process for Windows, 233-251
Hurricane watch, 72-79
Hyperlinks, 17, 62, 150, 178
Hypermedia, 17
Hypertext, 10, 17. *See also* HTML

I

Icon directives, 246-247
Icons, toolbar, 28, 37, 185-187
Identifying documents. *See* URLs
Images
embedded in HTML documents, 119-120
processing, 76-79
suppressing inline, 68
viewing, 72-79
IMG element in HTML, 119-120
Index searches (Gopher), 8
Information searches, 81-110
Information searching examples, 99-110
Information searching techniques, 96-99
Information services, 4
Information sources in this text, 192-196
Information structures, web and tree, 8
Inline images, suppressing, 68
Installing Chameleon, 208
Installing and configuring the server, 233-251
Installing HoTMetaL, 122-126
Installing Mosaic, 205-231
Installing Mosaic for Windows, 22-24, 217-222
Installing Mosaic for X Window, 223-226
Installing Trumpet Winsock, 215
Installing viewers, 228-231
Installing the Windows server, 234-236, 238, 243-247
Internet, 1-19
business applications for, 10-11
connecting to, 22-23, 40
direct TCP/IP access to, 22, 40
history of, 3
introduction to, 3-11
as a library, 4
management issues, 15-17
types of connection to, 205-206
what it does, 4
Internet access guidelines, 16
Internet Chameleon. *See* Chameleon
Internet Chameleon program group, 209
Internet (IP) addresses, 13-14

Internet newsgroups, 7, 83, 109-110
Internet provider setup program, 217-218
Internet Resource Meta-index, 50
Internet screen, text-based, 3
Internet starting points. *See* Starting points
Isolating the Internet computer, 16
Italics tags in HTML, 119

K

Key word searches of Internet database titles, 7-8
Key word searches in Veronica, 105-106

L

Learning by doing, 16
Library, Internet as, 4
Lists tags in HTML, 118
Load to Disk option, 73, 173
Local disk, saving a document to, 173
Logging into the server (Trumpet Winsock), 216
Login Settings dialog box (Chameleon), 211
Logons, recording unsuccessful, 16

M

Mail Document dialog box (Mosaic for X Window), 46
Main menu (Mosaic for Windows), 24-36, 170-180
Main menu (Mosaic for X Window), 180-184
Markup, 114, 120
Markup menu (HoTMetaL), 134-136
Menu items that are toggles, 27
Menus
    Mosaic, 169-184
    pop-out, 157-159
    top-level, 157
Messaging standards (MIME), 226
MIME (Multipurpose Internet Mail Extensions), 226
Modem settings for Chameleon, 207
Modem Settings dialog box (Chameleon), 211
Modems, 14400 baud, 13
Mosaic. *See* Mosaic for Windows; Mosaic for X Window (Unix)
Mosaic configuration file (Unix), 143, 163-166
Mosaic information files (Unix), 165-166
Mosaic for Windows
    Annotate menu, 31-32, 177
    Debugging info menu, 30
    downloading the files, 219-221
    Edit menu, 26-27, 171-172
    File menu, 25-26, 170-171
    finding viewers, 227
    getting a copy, 219
    getting started with, 21-37
    Gopher Servers menu, 34
    Help menu, 33, 35-36, 180
    History option, 32
    Home Page, 57
    installing, 217-222
    installing viewers, 228-229
    main menu, 24-36, 170-180
    Mosaic directory files, 221-222
    Navigate menu, 30-32, 175-177
    Online Documentation, 36
    Options menu, 27-29, 172-174
    Other Documents submenu, 35
    requirements, 218
    Starting Points menu, 33-35, 179
    toolbar, 37, 185-187
    User's Guide, 36
    World Wide Web Info submenu, 34
Mosaic for X Window (Unix)

Annotate menu, 49-51, 183-184
buttons, 52, 187-188
customizing, 226
downloading the files, 223-225
File menu, 42-46, 180-182
getting a copy, 223
getting started, 39-52
Help menu, 51, 184
installing, 223-226
installing viewers, 229-231
main menu, 41-51 180-184
Navigate menu, 47-49, 183
Options menu, 46-47, 182
testing, 40-41

Mosaic.ini file, 143-163
Annotations section, 156-157
comments in, 145
Document Caching section, 161
Font sections, 161-162
Hotlist section, 159-161
Mail section, 152
main section, 145-151
Main Window section, 152
Proxy Information section, 162
Services section, 153
Settings section, 152
statement syntax, 144-145
Suffixes section, 155-156, 227, 229
User Menu sections, 157-159
Viewers section, 153-155, 226-229

Movies
processing, 76-79
viewing, 77-79

Mpegplay viewer, 79
Mpg files, 79
MS-Windows. *See* Windows

N

Navigate menu (Mosaic for Windows), 30-32, 175-177
Navigate menu (Mosaic for X Window), 47-49, 183
NCSA Mosaic. *See* Mosaic for Windows; Mosaic for X Window (Unix)
NCSA Mosaic Home Page, 2, 23-24, 57, 115
HTML source code for, 114
internal view of, 27
testing, 23-24
URL for, 82
NCSA (National Center for Supercomputing Applications), 22, 40
Net News, 7
Net News URL, 83
Network ID, 13
Network Information Center (NIC), 13
Network node name, 13
News, accessing, 107-110
News reader, 7
News URL, general form for, 83
News.cso.uiuc.edu, 153
Newsgroups, 7, 83, 109-110
News.software newsgroup, 7
News.software.readers, 7
NNTP_Server variable in mosaic.ini file, 153
No Menus error, 166
Number variable in mosaic.ini file, 161

Online documentation for Mosaic, 162-164
Online Documentation option (Mosaic for Windows), 36
Online etiquette, 15
Open Ascii Styles dialog box (HoTMetaL), 135
Open Configuration File window (Chameleon), 209-210
Open dialog box (HoTMetaL), 127
Open Local Document dialog box (Mosaic for X Window), 43

Open template dialog box (HoTMetaL), 128
Open URL dialog box, 25
Opening screens. *See* Home Pages
Options menu (Mosaic for Windows), 27-29, 172-174
Options menu (Mosaic for X Window), 46-47, 182
Other Documents submenu (Mosaic for Windows), 35

P

Packet driver, 214
Paragraph marks in HTML, 117-118
Passwords, 15
Permanent vs. temporary configurations, 170
Personal Annotation window, 179
Personal annotations, 32, 178-179
Personal comments, 31, 49
Personal computer access to Internet, 12-13
Personal Home Pages, 87-90, 249 (*see also* Home pages)
  advantages of, 84-85
  creating, 86-91
  creating by cut and paste, 91
  sample, 86-91
  source code for, 86-91
Personal Menus window (Mosaic for Windows), 176
Plain text files as e-mail, 6
Point of interest, adding, 58-59
Pop-out menus, 157-159
Port Settings dialog box (Chameleon), 210
PPP (Point-to-Point Protocol), 13, 22-23, 40
Preformatted text tags in HTML, 118
President's public shedule, 71
Previewing a document, 172
Previewing in HoTMetaL, 129
Print Document dialog box (Mosaic for X Window), 44
Privacy issues, 16-17
Product information, 11
Protocols, communications, 12-13. *See also* Ftps
Public comments, 31, 51

Q

QUICKLIST URLs, 159-161
QUICKLISTS, 176-177

R

Radio station Home Page, 59-60
Radio stations, accessing, 59-60
Recording unsuccessful logons, 16
Rem keyword in mosaic.ini file, 145
Round List Bullets variable in mosaic.ini file, 150

S

Save Document dialog box (Mosaic for X Window), 45
Saving a document, 26
Saving a document to a local disk, 173
Screen display, modifying, 73-76
Searching for information, 81-110
Searching techniques, 96-99
  with Archie, 7-8
  examples of, 99-110
  structured, 99-102
  unstructured, 99, 103-108
Security issues, 15-16
Server configuration file, Unix version, 239-241
Server name variables in mosaic.ini file, 153

Server resource configuration file, 243-247
Servers, 17
configuring, 237-247
downloading the Unix version, 236-238
downloading the Windows version, 234-236
installing and configuring, 233-251
installing the Windows version, 234-236, 238, 243-247
shutting down, 248, 250-251
starting, 247-248
Service provider settings, 217-218
Show URLs variable in mosaic.ini file, 151
Shutting down the server, 248, 250-251
SLIP (Serial Line Internet Protocol), 13, 22-23, 40
SMTP_Server variable in mosaic.ini file, 153
Starting points, 49, 65
Starting Points Document, 104
Starting Points menu, 33-35, 96, 179
Starting the server, 247-248
Status bar, removing from the screen, 76
Status bar variable in mosaic.ini file, 151
Structured search techniques, 99-102
Surfing, 15
System name, 13

T

TCP/IP direct access to the Internet, 22, 40
TCP/IP protocol, 205, 214-216
TCP/IP services, 13
TCP/IP (Transmission Control Protocol/Internet Protocol), 12-13, 205-206
Technical support, 11-12
Telnet command, 7-8
Temporary vs. permanent configurations, 170
Testing NCSA Mosaic, 23-24, 40-41
Text files as e-mail, 6
Text formatting elements in HTML, 116-119
Text-based Internet screen, 3
TheCompany Home Page, creating, 138-140
Threads, news, 7
TITLE elements (in HTML), 116
Titles (document), removing from the screen, 74-75
Title/URL bar variable in mosaic.ini file, 151
Toggles (menu options), 27, 172
Tool tips, 185
Toolbar, 37, 185-187
removing from the screen, 73, 75
setting display of, 151
Toolbar variable in mosaic.ini file, 151
Top-level domain names, 14
Top-level menus, 157
Tour of Mosaic, 55-79
Travel information services, 4-5
Tree information structure, 8
Troubleshooting, 166-167
Trumpet Winsock, 205, 214-216
Two-directional Home Page, 95

U

Underlined text tags (in HTML), 119
Underlining of hyperlinks, setting, 150
University of Illinois at Urbana-Champaign Home Page, 94-95
Unix (*see also* Mosaic for X Window)
configuring Mosaic for X Windows, 163-166
downloading the server, 236-238
main server configuration file, 239-241

server startup script, 248
shutting down the server, 250-251
and TCP/IP, 13
Unstructured search techniques, 99, 103-108
Unsuccessful logons, recording, 16
URLs (Uniform Resource Locators), 25, 56
displaying, 151
explained, 82-84
ftp (file transfer protocol), 83
Gopher, 83
for NCSA Home Page, 82
news, 83
opening Unix, 43
parts of, 82
QUICKLIST, 159-161
removing from the screen, 74-75
Usenet News URL, 83
Usenet Newsgroups, 7, 108
UserDir directory, 244
User's Guide (Mosaic for Windows), 36

Veronica search tool, 8, 104-106
View menu (HoTMetaL), 132-134
Viewers, 56, 78, 226-231
Viewers and players, 226-231
Virtual Home Page, 92
Virtual Tourist sees North America, 101

WAIS (Wide Area Information System), 105, 107-108
Web information structure, 8
Web, The. *See* World Wide Web
What's New With NCSA Mosaic and the WWW, 64-65
White House search, 68-72
Window background, setting, 149
Window History dialog box (Mosaic for X Window), 48
Window menu (HoTMetaL), 138
Windows (*see also* Mosaic for Windows)
adding a Mosaic icon, 222
available command-line flags, 248
conventions for menu items, 25
Mosaic configuration file, 143-163
pull-down menu conventions, 25
TCP/IP for, 13
Windows Mosaic. *See* Mosaic for Windows
Windows server
downloading, 234-236
installing, 234-236, 238, 243-247
shutting down, 250
startup checklist, 247
Windows Server Home Page, 249
World Cup USA '94, 4
World Wide Web Info submenu (Mosaic for Windows), 34
World Wide Web screen, 10
World Wide Web (WWW or W3), 8-10, 62
WWW documents
adding to a Hotlist, 59
identifying and creating, 82
previewing, 129, 172
saving, 26
saving to a local disk, 173
WWW Info hyperlink, 100
WWW map of the world, 100
WWW servers, 56-58

X Window, configuring Mosaic for, 163-166
X Window Inside and Out, 226
X Window Mosaic. *See* Mosaic for X Window (Unix)

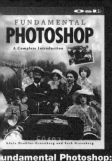

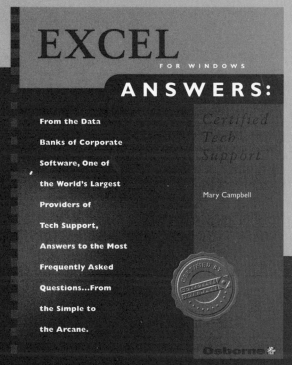

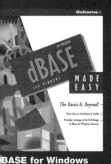

MAKE THE RIGHT CONNECTION

IT'S WHAT YOU KNOW THAT COUNTS. WITH INNOVATIVE BOOKS FROM LAN TIMES AND OSBORNE/MCGRAW-HILL, YOU'LL BE THE ONE IN DEMAND.

LAN TIMES
ENCYCLOPEDIA OF
NETWORKING
BY TOM SHELDON
AN AUTHORITATIVE
REFERENCE ON ALL
NETWORKING FACETS
AND TRENDS.
$39.95 U.S.A.
ISBN: 0-07-881965-2
AVAILABLE NOW

LAN TIMES
GUIDE TO SQL
BY JAMES R. GROFF AND
PAUL N. WEINBERG
$29.95 U.S.A.
ISBN: 0-07-882026-X

LAN TIMES E-MAIL
RESOURCE GUIDE
BY RICK DRUMMOND AND
NANCY COX
$29.95 U.S.A.
ISBN: 0-07-882052-9

LAN TIMES GUIDE
TO INTEROPERABILIT
BY TOM SHELDON
$29.95 U.S.A.
ISBN: 0-07-882043-X

BC640SL

Revolutionary Information on the Information REVOLUTION

The Business Week Guide to Global Investments Using Electronic Tools
by Robert Schwabach
Includes Three 3.5-Inch Disks
$39.95 U.S.A. ISBN: 0-07-882055-3

Alluring opportunities abound for the global investor. But avoiding investment land mines can be tricky business. The first release in the Business Week Library of Computing lets you master all the winning strategies. Everything is here—from analyzing and selecting the best companies, to tax planning, using investment software tools, and more. Disks include MetaStock, Windows On WallStreet, and Telescan, the leading investment analysis software.

The Business Week Guide to Multimedia Presentations Create Dynamic Presentations That Inspire
by Robert Lindstrom
Includes One CD-ROM
$39.95 U.S.A.
ISBN: 0-07-882057-X

The Internet Yellow Pages
by Harley Hahn
and Rick Stout
$27.95 U.S.A.
ISBN: 0-07-882023-5

BYTE's Mac Programmer's Cookbook
by Rob Terrell
Includes
One 3.5-Inch Disk
$29.95 U.S.A.
ISBN: 0-07-882062-6

Multimedia: Making It Work, Second Edition
by Tay Vaughan
Includes
One CD-ROM
$34.95 U.S.A.
ISBN: 0-07-882035-9

BC640SL

Yo Unix!

Innovative books from Open Computing and Osborne/McGraw-Hill

Open Computing's Guide to the Best Free Unix Utilities
by Jim Keough and Remon Lapid
Includes One CD-ROM
$34.95 U.S.A.
ISBN: 0-07-882046-4
Available now

Open Computing's Best UNIX Tips Ever
by Kenneth H. Rosen, Richard P. Rosinski, and Douglas A. Host
$29.95 U.S.A.
ISBN: 0-07-881924-5
Available now

Open Computing's Unix Unbound
by Harley Hahn
$27.95 U.S.A.
ISBN: 0-07-882050-2

Open Computing's Standard Unix API Functions
by Garrett Long
$39.95 U.S.A.
ISBN: 0-07-882051-0

BC640S

Think Fast PASSING LANE AHEAD

LOTUS NOTES

ANSWERS:

Certified Tech Support

From the Data Banks of Corporate Software, One of the World's Largest Providers of Tech Support, Answers to the Most Frequently Asked Questions...From the Simple to the Arcane.

Polly Kornblith

Osborne

Lotus Notes Answers: Certified Tech Support
by Polly Russell Kornblith
$16.95 U.S.A.
ISBN: 0-07-882055-3

What's the quickest route to tech support? Osborne's new Certified Tech Support series. Developed in conjunction with Corporate Software Inc., one of the largest providers of tech support fielding more than 200,000 calls a month, Osborne delivers the most authoritative question and answer books available anywhere. Speed up your computing and stay in the lead with answers to the most frequently asked end-user questions—from the simple to the arcane. And watch for more books in the series.

The Internet Yellow Pages
by Harley Hahn and Rick Stout
$27.95 U.S.A.
ISBN: 0-07-882023-5

Sound Blaster: The Official Book, Second Edition
by Peter M. Ridge, David Golden, Ivan Luk, Scott Sindorf, and Richard Heimlich
Includes One 3.5-Inch Disk
$34.95 U.S.A.
ISBN: 0-07-882000-6

Osborne Windows Programming Series
by Herbert Schildt, Chris H. Pappas, and William H. Murray, III
Vol. 1 - Programming Fundamentals
$39.95 U.S.A.
ISBN: 0-07-881990-3
Vol. 2 - General Purpose API Functions
$49.95 U.S.A.
ISBN: 0-07-881991-1
Vol. 3 - Special Purpose API Functions
$49.95 U.S.A.
ISBN: 0-07-881992-X

The Microsoft Access Handbook
by Mary Campbell
$27.95 U.S.A.
ISBN: 0-07-882014-6

BC640SL